THE
SPACING OF PLANETS:
The Solution to a 400-Year Mystery

Featuring

The New Fourth Law of Planetary Motion
The Comet S-L9 Collisions With Jupiter
How Earth Became Earth
Origin of the Universe:
Big Bang vs. Little Bangs

Alexander Alan Scarborough

Authors Choice Press
San Jose New York Lincoln Shanghai

The Spacing of Planets
The Solution to a 400-Year Mystery

Authors Choice Press
an imprint of iUniverse.com, Inc.

For information address:
iUniverse.com, Inc.
5220 S 16th, Ste. 200
Lincoln, NE 68512
www.iuniverse.com

Originally published by Ander (self-published)

ISBN: 0-595-15590-1

Printed in the United States of America

CONTENTS

If there is anything that can bind the heavenly mind of man to this dreary exile of our earthly home and can reconcile us with our fate so that one can enjoy living, -- then it is verily the enjoyment of the mathematical sciences and astronomy.

Johannes Kepler

PREFACE

The most fundamental mystery in science and in human imagination is the nature of our origins. Throughout civilization, humankind has been obsessed with trying to understand how and why everything came into being. This book builds on the solid foundation of understanding established by many great names from the past. Each of their discoveries dispelled established myths pertaining to our origins.

The works of Copernicus, Galileo, Kepler and Newton gave astronomers an understanding of the way our Solar System (SS) functions. In his quest for geometric solutions to planetary motions, Johannes Kepler (1571-1630) discovered the first Three Laws of Planetary Motion. Keenly aware of the Golden Ratio (GR) that later became recognized as Nature's Phi geometry, he recorded it as "a precious jewel." However, Kepler's attempts to solve the mystery of the geometric spacing of the six known planets proved futile.

In 1766 Titius of Wittenberg fitted these spacings into a mathematical formula that became known as Bode's Law (BL). Until 1995, the true significance of the BL spacing of planets remained a baffling mystery to anyone attempting to understand its meaning.

Why are the first eight planetary orbits spaced geometrically? Why don't the last two orbits comply with this geometry of the SS? Most scientists have given up on trying to explain BL, discounting it as a mere coincidence in numbers! But unlike pseudoscience, true science cannot discount data and facts simply because they are not understood. In reality, BL is the enigmatic key that is unlocking the mysteries of the origin, spacing, evolution and behavior of our planets!

In August of 1993, the significance of this crucial key to the origin of our SS was recognized by the author as simply a flawed version of the GR geometry "found flung broadcast throughout all Nature." In 1994 and 1995, the precise geometric relationship between the GR and BL was detailed to reveal the Fourth Law of Planetary Motion (FLPM) that had remained an enigma for 400 years after Kepler's initial recognition in 1595 of the geometric nature of the spacing of the six known planets.

This discovery attests to Einstein's belief in Nature's beautiful simplicity. In the perspective of the new Fourth Law, it is finally possible to explain (beyond reasonable doubt) why the first eight planetary orbits are geometrically spaced in close accord with BL, and why the last two planets do not obey BL!

Within the realm of this explanation lie the solutions to all anomalies of our SS. Now for the first time, the geometric, fiery origin of the SS, the spacing of planets and the subsequent evolution of all planets and planetary systems via internal nucleosynthesis (IN) can be explained within the scope (and by the means) of natural laws in a beautiful continuity of solid evidence. The FLPM/IN version is the only concept capable of making such claims and backing them with facts.

The ideas in this book are the culmination of 22 years of research into all past and current beliefs about our planetary origins. Not surprisingly some little-known, but accurate interpretations and a few brilliant ideas emerged from among the relevant hypotheses suggested throughout the pages of scientific history. Sadly, some of the most astute observations and accurate interpretations have lain buried beneath the onslaught of more popular, but erroneous beliefs.

For example, a number of name scientists, both past and present, have expressed their conviction that hydrocarbon fuels (gas, petroleum, coal) were not made from fossils. Yet the "fossil fuels" hypothesis has prevailed for some 160 years in spite of the powerful evidence that has accumulated against it. Even the crucial findings of William E. Logan (the originator) in the 1830s now are strong bits of evidence that argue *against* his erroneous conclusion that coal is a fossil fuel. His valuable discoveries proved essential in deriving and proving the new "energy fuels" theory of 1973, a concept that seems destined to displace the "fossil fuels" hypothesis before the turn of the century.

Strangely, the fallacy of the "fossil fuels" theory was the spark that eventually flamed into the discovery of the proposed Fourth Law of Planetary Motion. This book is a condensed version of these revolutionary ideas that have intermeshed precisely during the past 22 years. The proposed Fourth Law of Planetary Motion proved to be the final grand piece that ties everything together in the big jigsaw picture puzzle. In the words of Logan: My facts I now consider established beyond controversy.

However, this revolutionary concept yet remains dormant within the greater scientific community, a situation that created the reason and purpose of this book. Perhaps the future will grant a kinder fate than the past.

FOREWORD

During the past 22 years (1973-1995), a revolutionary concept of the origin of our Solar System (SS) and the evolution of its planets has been carefully put together. Like pieces of a jigsaw puzzle, a multitude of facts interlock precisely in place to reveal the sheer beauty of a more logical and provable concept of our planetary origins.

Throughout this revolutionary concept in planetology, a common thread connects and explains seemingly disparate observations of all planets and their planetary systems. That interconnecting thread is the key process know as internal nucleosynthesis (IN): the internal transition of energy to matter, a process common to all planets, spherical moons, our Sun and stars. In this IN perspective, each relevant discovery since the time of Copernicus can be interwoven into its fabric of supportive evidence.

The concise discussion in the first chapter explains the beautiful geometry of the SS that reveals how planets were placed in geometrically-spaced orbits. The next chapter describes how planets evolve through five common stages (from energy to inactive spheres). The third chapter explains how all planetary systems (atmospheres, land, water, fuels, etc.) evolve from energy. The principles involved are best illustrated in the detailed discussion of the evolution of hydrocarbon fuels (gas, oil, coal) by means of natural laws.

Through an intermeshed continuity of scientific evidence, the revolutionary FLPM/IN concept described herein has come full circle from the energy fuels theory (EFT) of 1973 to the Fourth Law of Planetary Motion (FLPM) of 1995.

Before beginning, it would be appropriate to review the six firm rules for truth in science:
1. Do not violate the laws of physics and chemistry.
2. Do not rely on hypotheses, speculations or assumptions.
3. Have a continuity in which each stage connects with the preceding and succeeding stages of the grand scenario.
4. Every relevant fact must interlock precisely in place.
5. All anomalies must be factually explainable.
6. The mathematics must be accurate, and correctly interpreted.
   These rules serve as strict guidelines for the intermeshing concepts of the planetary origins described in this book.

The reader should note the ten illustrations grouped together, beginning on page 25, for easier references when comparing the intermeshing graphs and pictures. Separate sets suitable for framing are available from the author through the publisher.

5

When we try to pick out anything by itself, we find it is hitched to everything else in the Universe.

John Muir

...scientific certainty comes from interpreting ambiguous results within an order scientists themselves impose. Only by coming to grips with this conclusion can we understand this powerful yet fallible creation that we call science.

The Golem
by
Harry Collins & Trevor Pinch

CHAPTER I

THE SPACING OF PLANETS

HISTORY'S FIRM FOUNDATION

In 1781 Sir William Herschel, while observing the constellation Gemini, noted that its brightest star was moving. It had to be a comet. Herschel could not believe that he had discovered an eighth planet; everybody knew that only seven planets existed in the Solar System (SS), seven being a magic number. Newton's laws enabled mathematicians to determine that the orbit was of a circular pattern, thus identifying the star as the eighth planet, Uranus. By the mid-nineteenth century, most astronomers had accepted this name, as suggested by the German Johann E. Bode.

It was appropriate that Bode should have named the new planet, because he had predicted its existence in the correct orbit before its discovery. Bode had rediscovered a mathematical relationship first observed by Titius of Wittenberg in 1766. Director of the Berlin Observatory and author of a vast catalogue of star positions issued in 1801, Bode is best known simply for popularizing this mathematical relationship:

If to each number in the series 0, 3, 6, 12, 24, 48, 96, 192, one adds 4, then the series becomes 4, 7, 10, 16, 28, 52, 100, 196. When these numbers are divided by 10, the resulting figures are the approximate distances of the respective planets from the Sun: Mercury = 0.4 AU, Venus = 0.7 AU, Earth = 1.0 AU, Mars = 1.6 AU, Asteroids = 2.8 AU, Jupiter = 5.2 AU, Saturn = 10 AU. The general formula is $D = 0.4 + 0.1X$, where X is 0, 3, 6, 12, 24, 48, 96, 192, (the original series above).

This relationship is still known as Bode's Law (BL). At the time it was popularized, no planet was known to exist in the orbits at 2.8 AU and 19.6 AU. Herschel's discovery of Uranus in the orbit at 19.6 AU, as predicted by this law, spurred the search and discovery of Ceres in the orbit at 2.8 AU. However, when Leverrier discovered Neptune in 1846, it was found in a position much closer to the Sun than that predicted by BL. Since then, the significance of this relationship, or law, of the spacing of planets has remained a mystery wrapped in an enigma, conveniently discarded by many scientists as a quaint coincidence in numbers.

However, in true science, such data cannot be tossed aside because its significance is not understood at the time. Although BL was not known in his time, Johannes Kepler (1571-1630), perhaps history's greatest astronomer, realized that the spacing of the six known planets was of a geometric nature. While successful in establishing the First Three Laws of Planetary Motion by solving their relevant mathematics, Kepler failed in many attempts to solve the geometric spacing of the planets -- a potential Fourth Law of Planetary Motion.

Although on the right track, Kepler somehow missed the right train of thought. He wrote, "Geometry has two great treasures: one is the theorem of Pythagoras; the other, the division of a line into extreme and mean ratio [Golden Ratio (GR)]. The first we may compare to a measure of gold; the second we may name a precious jewel." The pervasiveness of the GR was expressed best by C. Arthur Coan: "Nature uses this as one of her most indispensable measuring rods, absolutely reliable, yet never without variety, producing perfect stability of purpose without the slightest risk of monotony...We shall find it flung broadcast throughout all Nature."

In his book, *The Divine Proportion*, H.E. Huntley wrote, "It was suggested in the early days of the present century that the Greek letter ϕ (Phi) -- the initial letter of Phidias's name -- should be adopted to designate the golden ratio. The ubiquity of Phi in mathematics aroused the interest of many mathematicians in the Middle Ages and during the Renaissance. In 1509 there was published a dissertation by Luca Pacioli, *De Divina Proportione*, which was illustrated by Leonardo da Vinci. Reproduced in a handsome edition in 1956, it is a fascinating compendium of Phi's appearance in various plain and solid figures. We shall in following chapters come across many examples of the appearance of Phi in unexpected places." The examples are indeed numerous.

THE PROPOSED FOURTH LAW OF PLANETARY MOTION

Had Kepler recognized the true significance of the precious jewel that actually was in use before 500 B.C. and that now seems destined to become the universal constant Phi, he might have utilized it as a basic tool for understanding the orbital velocity/distance/spacing relationship of each of the six known planets. When this was finally done in 1980 by the author, a smooth curve of the current (BL) Phi relationship of these properties of all planets was drawn. Here, Phi would prove to be the key to understanding BL, and eventually to unraveling the mystery of the geometric origin of our SS. However, its full significance was not to be recognized until 1994-1995, some fifteen years later.

In 1994 the dawn began to break: The current spacing of planets could not possibly be their original orbits of some five billion or more years ago; everything in Nature forever changes, nothing remains forever as it once began; gravity and time exact their toll on all of Nature's systems. BL had to be simply a flawed version of the original perfect spacing of the planets (in accord with Nature's universal Phi). Anyone's original assumption that orbits remain forever stable was dead wrong -- an actual stumbling block that had tripped too many theoreticians. The planetary orbits simply had been displaced from their original Phi orbits to their current positions, and will continue to be displaced by gravity forces throughout future eons. Thus, both past and future positions of the planetary orbits of our SS are calculable -- a conclusion based on Phi mathematics and BL, and strongly supported by much scientific evidence, as we shall see further along in this book.

In the concluding geometric solution (1995) to the spacing of planets, the smallest figure (0.618034) related to Phi (actually the reciprocal of Phi) represents the original distance of the closest planet (Mercury) from the Sun. The original distance of each planet thereafter was calculated by the formula, $D = Phi \times PO$, in which Phi is 1.618034 and PO is the distance (AU) of the previous orbit from the Sun. Through utilization of Kepler's Third Law of Planetary Motion ($D^3 = T^2$), the original orbital velocity of each planet was calculated. These sets of Phi coordinates were plotted on the same GR scale used for the BL coordinates fifteen years earlier. The result was a curve almost identical to the 1980 BL-GR curve.

Upon superimposing the two curves (BL and GR), certain conclusions became obvious:

1. BL is simply a flawed version of the original geometric GR (Phi) spacing of the planets.

2. The geometric spacing of planets is the crucial key to understanding the true scientific nature of our planetary origins.

3. The centuries-old enigmatic spacing of the planets finally has a provable solution based on mathematics, sound logic and strong supportive evidence.

4. *The gravity-induced displacement of any planetary orbit since the geometric origin of the SS is determined by the difference between its current (BL) position and its original Phi-spaced position. The original velocity of each planet is determined by Kepler's Third Law of Planetary Motion: $(D^3 = T^2)$.

*(A plausible Fourth Law of Planetary Motion)

5. The future displacement, velocity changes and the fate of each planet of the SS is predictable mathematically.

Here was geometry at its beautiful best -- beauty best defined by Richard Jefferies: "The hours when we are absorbed by beauty are the only hours when we really live...These are the only hours that absorb the soul and fill it with beauty. This is real life, and all else is illusion, or mere endurance."

With understanding, the concept grows in beauty and simplicity. During the past 22 years, much supportive evidence favoring this radically different perspective of planetary origins has accumulated in the scientific literature and in the author's files. In this book, three original sets of interlocking geometric diagrams, each set corroborating the other two and mingled in a pictorial history, explain how and why planets arrived at their current orbital positions in our SS. By combining this plausible Fourth Law of Planetary Motion with the author's previous writings on the evolution of planets via internal nucleosynthesis (the internal transition of energy into matter) in this revolutionary concept, reasonable solutions to a long list of anomalies of our SS become feasible.

In his *Principles of Philosophy* Descartes defined Earth's interior as being "Sun-like". Later, Buffon, author of the 36-volume *Natural History*, speculated that "fragments of the Sun must have been knocked off into space then came together to form spheres revolving in the same direction and in the same plane. Each of them became a planet turning on its own axis, flattened at the poles. And satellites were thrown out." (Ref: *The Discoverers*, Daniel Boorstein).

Considering the limited knowledge of the time, these two beliefs are amazing in how close they came to an elementary understanding of our planetary origins long before the relationship between nuclear energy and atoms and the processes of nucleosynthesis and polymerization were recognized.

Briefly stated, Einstein's formula, $E = mc^2$ revealed the relationship between nuclear energy and matter, making it obvious that all atomic matter

10

formed from nuclear energy. Such transformations must occur, of necessity, under extreme conditions of high temperature and pressure within sources of nuclear energy. Various combinations of these severe conditions determine the types and quantities of elements produced via nucleosynthesis within each planet and moon.

Elements ranging from hydrogen to iron have been detected in our Sun. Scientists have concluded that no elements heavier than iron can be found therein. However, upon encapsulation of such Sun-like nuclear energy, more severe conditions of temperature and pressure slowly build within the mass. For example, as the initial gaseous atmosphere and later the liquid and crustal formations gradually encapsulated the hot energy mass destined to be named Earth, these internal conditions grew ever more severe, thereby creating heavier elements -- eventually producing uranium, the last and heaviest of a long list of elements in the formation of our planet's ever-thickening crust (see Figure 1).

Thus the production of heavy elements comprising planetary crusts occurs in encapsulated energy masses as well as in supernovae and atomic bomb explosions. (Note: Heavy elements 99 and 100 were produced in the hydrogen bomb explosion). In this perspective, each planet is a self-sustaining entity: Via the natural processes of nucleosynthesis and polymerization, its mass of nuclear energy continually undergoes transformation into atomic elements that combine as molecules comprising all matter of each planet as the sphere progresses through various stages of evolution from energy to gaseous to rocky to inactive body.

If, as will be explained in Chapter II, planets are indeed self-sustaining entities transforming via nucleosynthesis from nuclear energy masses to gaseous to rocky to inactive spheres in accordance with the laws of thermodynamics and as functions of size (see Figure 2), how were they geometrically spaced in orbits with such clockwork precision?

HOW PLANETS WERE PLACED IN GEOMETRICALLY SPACED ORBITS

The First Three Laws of Planetary Motion were discovered by Johannes Kepler in the early seventeenth century. Dealing with the precision of the SS, these laws describe the elliptical orbits and precise clockwork of its planets:

1. The orbit of a planet is an ellipse, of which one focus is the Sun.

2. For each planet a straight line joining it to the Sun sweeps over an equal area in equal time.

3. For each planet the cube of its distance (AU) from the Sun is equal to the square of its time (in years) of revolution ($D^3 = T^2$).

These laws reveal *the absolute necessity of a great momentum of a smaller, faster mass interacting with the Sun as it speeds past the huge fireball to create the elliptical orbits with geometric clockwork precision.*

Binary star systems in which the two stars interact continuously are known to exist in large numbers. Imagine a potential binary star system in which the smaller, faster mass zips past the larger mass *at a specific distance apart.* If the two masses are of *proper relative sizes and have proper relative velocities,* the result will be a breakup of the smaller mass into fragments that split off at points determined by Nature's ubiquitous GR (Phi) geometry. With this special scenario in mind, one begins to realize *why* *and how* the planets of our SS were placed precisely in geometrically spaced orbits. By combining this realization with the mysterious BL for the current spacing of planets, one can derive a plausible Fourth Law of Planetary Motion in which significant changes in the orbits and velocities of planets are revealed and future changes predictable.

The fiery origin of the SS is illustrated in the painting, *Birth of the SS* (Figure 3). This artistic rendition of the manner in which the origin and geometric spacing of planets occurred allows an easier understanding of the geometry and the geometric forces to be illustrated in several diagrams later in this book. In it we see the breakout of Jupiter and its moons, far from the inner nebulous planetary masses orbiting the Sun in the lower left corner.

In every galaxy throughout the Universe untold numbers of binary star systems have been and are being created when two fiery masses, energized by Nature's more powerful explosions (Little Bangs), lock in orbit around each other. In *rare and special cases,* two such masses will meet the precise prerequisites for creating planetary systems: proper relative sizes and velocities and proper distance apart to effect the fragmentary breakup of the smaller, faster mass as it swings by and beyond the larger mass *at a decelerating pace.*

In Figure 4 the BL curve shows the current relationship between the orbital velocities and distances of planets when plotted in the Phi scale. What is the significance of BL, a formula for the geometric spacing of the first eight planets (including the Asteroids)? And why do the last two planets, Neptune and Pluto, fail to comply with this law? Is the spacing of planets really a crucial key to understanding and proving the origin of our SS? The answers to these questions reveal a beautiful continuity of evidence that offers golden opportunities to understand the origin and evolution of planets.

This nearly perfect AU-OV-BL curve, drawn in 1980, lulled me into assuming this to be the original planetary orbits and velocities that had not changed since the original layout of the SS some five billion years ago. The assumption was wrong. Years passed before my realization that orbital positions and velocities do change, and that nothing can stay in orbit forever. The original positions obviously had to comply with Nature's ubiquitous Phi law of spacing, and the corresponding velocities had to comply with Kepler's Third Law of Planetary Motion! The BL curve for the current positions and velocities of planets is simply a flawed version of Nature's ubiquitous GR law of geometric spacing!

In July, 1994, the original relative motion diagram (SS-6 of 1980) depicting the geometric origin of the SS was modified to reflect a comparison of the BL data versus the GR data on the same geometric drawing utilizing the GR Triangle. Diagram SS-6-R4i (Figure 5) for the inner planets shows the relative motions of the Sun and the smaller mass during the origin of the SS. As the Sun moved at _a steady pace_ to the right along the X-axis, the smaller energy mass (SEM) moved at a 26½° angle up the hypotenuse of the Architect Triangle (AT) at _a decelerating pace_, releasing a fiery fragment at each Phi point along the way. The GR distance of 0.618 AU between the two properly proportioned masses was a crucial factor in the initial break-off from the SEM of a fiery fragment that was destined to evolve into Planet Mercury.

Diagram SS-6-R4o (Figure 6) is an expansion of the previous diagram to include the Phi release points and paths taken by all ten nebulous masses (including the Asteroids). For example, just as the Sun reached the right end of the X-axis, Pluto, the last of the ten masses, was pulled simultaneously into the final orbit at the top of the Y-axis. Had an observer ridden above the Sun during the time each nebulous planetary mass was pulled into orbit, the path of the smaller mass along the hypotenuse of the AT would have appeared to be the smoothly curved GR line extending from S (Sun) to P (Pluto), _an optical illusion caused by the relative motions_ of the two masses.

The Architect Triangle is established via the standard geometric method of dividing a line into extreme and mean ratio. Contained within the Triangle are all the figures pertaining to the GR, the Fibonacci Series, the Logarithmic Spiral, and perhaps to all the geometric forces of the Universe. The justification for its epithet was realized when its angles (26½° and 63½°) furnished a solid basis for construction of the seven SS Diagrams that intermeshed so precisely during the geometric solution to the origin of the spacing of the planets. In the author's opinion, a yet unknown connection exists between the AT and the superstrings theory of universal origins. (More on this later).

The warped curve identified as BL in Figure 6 is a plot of the current positions of the planets. Thus, one can conclude that BL of spacing is simply a flawed version of the original GR law of geometric spacing of the planets -- spacing warped by the forces of gravity over a very long time frame of some five billion years.

As the smaller mass swung at a 12½° (to be explained later) around the Sun, several other things happened. The passing SEM pushed the Sun's equatorial gases to a faster pace than its remaining mass and at an angle of 7° to the plane of the ecliptic. Thus began the erratic distribution of different degrees of inclination of the planets to the ecliptic, each effected by the combined forces of gravity of the Sun and of the preceding planetary mass(es). Mercury's angle of 7°, the same as that of the Sun's equatorial gases, is an important clue in verification of this concept.

At its Phi point, the Venus fragment broke from the upper part of the trailing edge of the rotating (counterclockwise) SEM, thus accounting, via a reverse gear-type action, for its slow, unique spin in the clockwise direction. Thereafter, at each successive Phi point along the way, gravity forces of the Sun and the previous planet(s) combined to pull the next fragmentary mass into orbit, each mass rotating in the counterclockwise direction (viewed from above). The layout procedure is best illustrated in the diagrams shown in Figures 7 and 8. This second set of diagrams (modified versions of the original 1980 diagrams) show _how_ each planetary mass was pulled into orbit by the combined forces of gravity of the Sun and the preceding planet in accordance with Nature's ubiquitous GR. Note that no attempts to include the chaotic effects of the gravities of secondary planets in each case have been made.

In both diagrams, the Sun moves along the X-axis as the SEM moves up the Y'-axis. The geometry of the forces of gravity pulling the _inner_ planetary masses (Figure 7) into their respective orbits furnishes corroborating evidence of the manner in which the clockwork precision described in Kepler's First Three Laws of Planetary Motion _came into being._ As each newly orbited fragment reached the tangent line (T or T'), its gravity acted in conjunction with the strong gravity of the Sun to pull the next mass into orbit, _thereby simultaneously varying the angle of inclination to the ecliptic for each planet-to-be._

The second diagram (Figure 8) of this set is expanded to show the geometry of the gravity forces acting on the outer planetary masses. As each mass reached the tangent line (T or T') during the fiery layout of the SS, the next nebulous mass was pulled into orbit at a different angle of inclination to the ecliptic. If the original angle of 7° had been 0°, all planetary orbits would have been in the same plane.

In both Figures 7 and 8, Y' can be made to coincide with Y at 90° by subtracting 14° from the baseline of 26½° to form a new baseline of 12½°. This shifts the exit points of all planets from Y' to Y and moves the T line down to T' to form a 39° angle with X. The significance of these moves that create a better balanced Diagram is that the angle of diversion of the SEM around the Sun actually was only 12½° instead of 26½°. While this interpretation is somewhat tenuous, it affects neither the conclusions drawn from the other Diagrams nor the relationships among the sets of Diagrams.

The elliptical shapes of planetary orbits can result only from the rapid velocity of the smaller mass from which the nebulous planets were placed in their original Phi orbits. The multitude of known binary star systems and the scarcity of known planetary systems provide strong evidence that twin-star systems _very rarely meet the prerequisites of Nature's ubiquitous Phi geometry to become a SS like ours_. But whenever one does, energy fragments of the smaller, faster mass will break off at Nature's GR points along the way, all in compliance with the Phi geometry of the gravity forces acting at those points.

The decision to draw the GR curve in the GR scale was made in early 1995 (see Figure 9). Through use of the GR spacing of planets and Kepler's Third Law, the original orbital velocities were calculated for every planet (including the Asteroids). Finally, the full dawn began to break: the GR curve, when plotted in the GR scale, _reveals the original positions and velocities of the nebulous planets during the fiery birth of the SS._ Since the 1995 GR (Phi) curve looked almost identical to the BL curve of 1980, why not superimpose the two curves to determine and evaluate the differences?

When the two curves are superimposed, they almost become one (Figure 10). However, some significant differences are noted for each and every planet. Of the ten original orbits, eight have moved closer to the Sun, while Uranus and Neptune have moved slightly away. Changes in the velocities correlate with changes in orbital positions in compliance with the Third Law. Uranus and Neptune were the only two planets showing negative results: slight losses in velocity correlating with slight increases in distances from the Sun. Each of the other eight planets (including the Asteroids) shows a significant increase in velocity that correlates with its decrease in distance from the Sun. If a planet (or asteroidal matter) does exist beyond Pluto, its original distance and velocity calculate to 76.01 AU and 1.06 miles per second, respectively.

Mercury, the innermost planet, has _gained_ the most velocity (6.18 mi/sec) while being among the planets _losing_ the highest percentage of its distance from the Sun. These increases in velocity and decreases in distance appear to be proportional to the mass/distance relationship of each planet and to the gravity effects of the preceding mass(es). Uranus and Neptune are examples of very distant planets, each with sufficient mass and distance to resist the inward pull of the combined gravities of the Sun and the inward planets; they are gradually drifting away while slowly decreasing in orbital velocity.

Contrary-wise, Pluto is so small that is has been, and still is, greatly affected by these gravities, in spite of its greater distance. This planet's tiny size permits it to be tossed around (relatively speaking) by any and all gravity forces within range, thus accounting for its highly irregular orbit and its decrease of 4.68 AU from its original Phi position at 46.98 AU. At this rate, Pluto's potential for collision with another planet in the far distant future seems real.

Besides Pluto, small significant differences between the superimposed curves occur at Uranus, Mars and the Asteroids. The large mass and distance of Uranus and the small mass of Mars account for their behavior, while the discrepancies of the Asteroids can be attributed to their vulnerable position proportionately balanced between the two huge masses of Jupiter (the geometric mean of the SS) and the Sun. Apparently, the original Asteroids mass broke into three main bodies, pulled apart by the gravity forces of the two larger masses. And exactly as Olbers concluded, the Asteroids disintegrated, exploded. Some 30,000 planetesimal remnants remain today in three distinct orbital paths, each path beautifully spaced in precise geometric configuration in the SS.

Looking at the superimposed curves, one can conclude once again that the BL spacing of planets is simply a flawed version of Nature's original GR spacing. This third set of diagrams correlates precisely with the results of the first pair of diagrams showing *the relative motions of the two masses* of the potential binary star system during the original geometric layout of the SS some five billion years ago and the changes in planetary orbits that have occurred since then.

Developed during a time frame of 15 years (1980-1995), the seven geometric diagrams, each interlocking in corroborative support of the others, have come full circle to reveal the true beauty of the geometric origin of our SS. The enigmatic BL of spacing of the planets finally has been exposed as a flawed version of the original Phi spacing of the planets. In full compliance with the laws of thermodynamics, the fiery energy masses slowly evolve via the processes of nucleosynthesis and polymerization through the various stages of evolution to end up eventually as cold, inactive spheres (see Chapter II).

From this pictorial history with its seven intertwined geometric diagrams, a plausible Fourth Law of Planetary Motion can be restated with a greater degree of confidence:

"The gravity-induced displacement of any planetary orbit since the geometric origin of the SS is determined by the difference between its current (BL) position and its original Phi position. The original velocity of each planet is determined by Kepler's Third Law of Planetary Motion ($D^3 = T^2$)."

Any relevant anomaly can be explained better in this new FLPM/IN concept than in any known theory of planetary origins. All are interwoven with a common thread throughout this revolutionary concept. While this book touches upon some two dozen anomalies, many remain beyond the realm of its present objective. All of them offer numerous opportunities to pursue a true understanding of the origins and evolution of planetary and universal systems.

HOW STABLE IS THE SOLAR SYSTEM?

Questions have been raised about the stability of the SS. Is the particular distribution of the planets just one of many possible stable arrangements? Or is it the one arrangement that survives because it happens to result in a stable SS?

Does the SS have room for an extra planet or two without disturbing its apparent equanimity? While such questions remain largely unresolved, their answers do appear to reside in the Fourth Law of Planetary Motion that is based on the Phi geometry of the SS.

The Phi geometry of the original positions of the planets gives the impression of solid stability. However, when this Phi curve of the original orbits is superimposed on the curve of the current spacing of the planets, the answers to questions on SS stability become obvious.

The chaotic displacements of the planets, while relatively small for all except Pluto, give a clear picture of the changes that have occurred since the origin of the SS. Each and every planet (including the Asteroids) has undergone small chaotic changes in both its orbit and velocity. For example, the superimposed AU-OV curves (Figure 10), reveal that Planet Earth has moved 38.2% closer to the Sun, while gaining 26.9% velocity. Similar changes for each and every planetary mass are recorded in the graph's tabulated data. From these changes, one can interpret the past and predict the future changes in each sphere's orbit.

This geometry clearly shows that there is no room to put another planet or two between the Sun and Pluto, the outermost planet. However, if one does not exist beyond Pluto, plenty of room is there for another planet -- but perhaps insufficient gravity precludes that possibility.

DO OTHER SOLAR SYSTEMS EXIST?

On June 11, 1994, the Hubble Space Telescope imaged the tiniest star ever recorded. Identified as Gliese 623b, the star is one-tenth as massive as the Sun and only one sixty-thousandth as bright. This diminutive mass lies 25 light-years from Earth, but shines too faintly and lies too close to a bigger, brighter companion (Gliese 623a) to be detected by ground-based telescopes. If placed as close to Earth as our Sun, the tiny star would have only eight times the brightness of a full moon. Astronomers estimate that the dim star lies some 320 million kilometers (2.14 AU) from its large companion and takes an estimated four years to orbit it.

The relationship between Gliese 623a and its tiny orbiting mass is governed precisely by the Four Laws of Planetary Motion. Astonomers know that the orbit has an elliptical shape and will sweep out equal areas in equal time while moving at accelerating and decelerating velocities in its elliptical orbit. In compliance with the Third Law, the cube of its distance from the larger mass will equal the square of its time of revolution. At this point in time, more information will be needed before the Fourth Law can reveal any displacement figures and the fate of the tiny star.

The establishment of the exact time of revolution enables scientists to calculate accurately the distance between the two masses and their distance from Earth. If the system is shown to be in compliance with the Four Laws of Planetary Motion, it might be a SS made in the same manner as ours. But unless smaller masses are found in other orbits, astronomers must conclude that the smaller mass failed to meet the precise specifications of the Phi geometry of the Fourth Law, and thus did not break up into smaller masses in the same manner as our planets did. Rather, the two masses took the usual route of remaining a binary star system, albeit an odd couple type of system.

The dimness of the small star leads one to suspect it of being in the last stage as a star, and is now entering the initial phase of the first stage of cooling and evolving into a gaseous planet in full accord with energy-to-matter laws; i.e., natural laws of physics, thermodynamics, chemistry, etc.

The big question then becomes: Is this the first recorded picture of a SS in its nebular stage of evolution from energy to planet? If so, it will rank among the most significant and exciting discoveries in the history of astronomy -- one that can lead to acceptable, even provable, solutions to all planetary origins.

THE AGES OF STARS AND GALAXIES

The conflict between the ages of the oldest stars versus the age of a younger Universe may force scientists to rethink the standard theory of the Universe. Observations by the Hubble Space Telescope imply that under the standard theory, the Universe would be about 9.5 billion years old. But scientists are confident that the oldest stars are at least 12 billion years old.

The ages of stars have been extensively studied; they cannot be lowered without resorting to fundamental changes in scientific understanding of particle physics, according to astronomer Nial Tanvir of Cambridge University in England. Tanvir recently concluded that the standard assumptions about the cosmos might be wrong.

While such differences in ages cannot be understood within the realm of current beliefs about universal origins (the Big Bang), the problem simply does not exist in the perspective of the Little Bangs theory (1980) in which energy is continuously created at the spherical perimeter of the Universe during its speed-of-light expansion. Subsequently, this energy serves as the source for the evolution of new systems: quasars, galaxies, stars, planets, moons, comets, asteroids, etc., all evolving in full accord with natural laws, the Phi geometry and the mathematics of the Universe.

"When we try to pick out anything by itself, we find it is hitched to everything else in the Universe," stated John Muir, an astute observer indeed.

In December 1995, the Hubble Space Telescope focused on a tiny patch of sky near the handle of the Big Dipper for 10 consecutive days, recording a brilliant picture that reaches far deeper into space than ever before. A bewildering array of galaxies were recorded from a composite of exposures from ultra-violet, blue, red and infrared emissions. The combined color image appears to show galaxies of all ages. In the current Big Bang perspective, the most distant ones were photographed as they looked when the Universe was only about one billion years old.

About 1,500 galaxies, many only one four-billionth as bright as the dimmest light the unaided human eye can see, can be seen in the spectacular Deep-Field picture. The galaxies are stacked up against one another, so the real challenge is to disentangle them. A high priority is to determine the distances from Earth for as many of the galaxies as possible. Since distant galaxies move away from each other faster than nearby galaxies, astronomers can determine within reason each one's distance from Earth.

According to the FLPM/IN concept, the most distant galaxies will be the most recent ones created at the expanding perimeter of the spherical Universe, and will be moving at nearly the speed of light (details in Chapter V).

Thus to fathom universal and planetary origins, mankind must learn their mathematical relationships and *their continuity in the chain of evolution from energy to all forms of matter. The proposed Fourth Law of Planetary Motion detailing the beautiful geometric spacing of the nebulous energy masses that subsequently and continuously evolve via nucleosynthesis and polymerization as self-sustaining planetary entities brings this conclusion sharply into focus.*

TWO MORE EXTRASOLAR PLANETS?

Since the writing of the first four chapters of this book, the discovery of two more extrasolar planets have appeared on the scene. At a meeting of the American Astronomical Society in San Antonio in January 1996, Geoffrey W. Marcy and R. Paul Butler presented a paper describing the discovery of two unseen planets, each orbiting a nearby star. One of the planets lies apparently at the right location from its parent star for liquid water to exist on its surface. The second planet might contain liquid water, but only in its atmosphere.

The astronomers discovered the two planetary masses around sunlike stars -- 70 Virginis in the constellation Virgo and 47 Ursae Majoris in the Ursa Major, also known as the Big Dipper. Although both stars are visible to the naked eye, the planetary masses are too small, and thus too faint, to be seen against the glare of their parent suns. So the researchers had to use the indirect technique of measuring small shifts in wavelengths of light emitted by the parent stars to find evidence of the existence of the orbiting masses.

Marcy and Butler monitored the motion of the stars with a spectrograph mounted on a 120-inch telescope. A computer analysis revealed that light emitted by the two stars appears alternately redder and bluer, indicating that they move back and forth along the line of sight to Earth. In each case, the tell-tale wobbles describe a nearly perfect sine curve-- a motion so periodic that only an unseen object pulling the star toward and away from Earth can account for it.

These discoveries, coming on the heels of the 51 Pegasi finding, are convincing evidence to scientists that planets are not rare after all. The planetary mass orbiting Ursae Majoris is about three and one-half times that of Jupiter. Circling its sun at about twice the distance of Earth from the Sun, the planet takes roughly three years to complete one revolution. Its surface temperature is estimated at -90°C.

In contrast, the sphere circling 70 Virginis has a mass about eight times that of Jupiter. Its orbit lies, on average, less than half Earth's distance from the Sun. Its surface temperature is about 83°C. Conceivably, the planetary sphere could have rain or bodies of water. However, assuming that it has a solid surface, its enormous gravity and high pressure would be crushing.

Further, the orbit of this planet is highly elliptical. Because of its strong gravity, a massive planet on an elliptical path tends to de-stabilize the orbits of nearby planets. Thus, the 70 Virginis is unlikely to possess an array of orbiting spheres similar to our SS.

There are other clues pointing to the origin and true nature of the two planetary spheres. First, the process of planet formation in which material accumulates from planetesimals and/or from a dusty disk rotating around a star does not permit a planet to have an elliptical orbit. Such elliptical orbits are permitted only in binary systems in which explosive momentum is created -- as happens in the FLPM/IN concept.

Secondly, the 51 Pegasi planetary sphere is much too close to its Sun to have formed from a gaseous dust-cloud that either would have burned away or fallen into its parent sun. Further, it is difficult to imagine such a fast moving target being bombarded with in-falling planetesimals that add substance and heat to its mass, as is hypothesized for planets of our SS, a vital process for building planets in the perspective of the BB theory.

To understand the true origin and nature of the three new solar systems, one must consider the relationships discussed in Chapter I: the relative sizes of each pair, their relative distance apart and their relative velocities. In this perspective, the logical conclusion is that all three of the new systems are representative of peculiar Mutt-and-Jeff pairings of binary star systems.

The large sizes of the two new planetary masses, each circling its parent star, indicate that they are in either the first stage (energy masses) or the second stage (gaseous) of evolution. Because of their extra large sizes, any betting odds would have to favor the first stage of evolution. The smaller 51 Pegasi mass, being only 0.6 the mass of Jupiter, could well be in the rocky stage of evolution. This statement is based on the assumption that it is no younger than our SS planets and perhaps somewhat older.

All three are fine examples of solar systems that failed to meet the precise prerequisites of the Phi geometry that proved so essential in the geometric spacing of the planets of our SS. Thus, it appears that each of the three new discoveries is destined to remain a binary system rather than a geometrically-spaced multi-planet SS like ours. (For relative locations of planets, see p. 26).

The full significance of the three discoveries may not reach fruition for many years. While presenting exciting evidence supportive of the FLPM/IN concept, they argue strongly against the Accretion Disk hypothesis and subsequently against its well-entrenched ally: the Big Bang theory.

THE QUANTUM REVOLUTION

The new theory of quantum mechanics came into existence in 1925, giving scientists the first comprehensive formulation with which to pry open the secrets of the atom. By 1926, quantum physicists had developed a mathematical description of the hydrogen atom, whose properties could now be explained by pure mathematics. By 1930, they were claiming that chemistry, simply put, is "applied physics."

Quantum theory is a direct opposite of Einstein's general theory of relativity in which stars and galaxies are held together by the smooth fabric of space and time. By contrast, in the quantum theory subatomic particles are held together by particle-like forces on the empty stage of space-time.

The quantum revolution negated all efforts at a geometric understanding of forces for more than 50 years. In this perspective, the material Universe consists of the 92 atomic elements with which all the known forms of matter can be built. These atoms consist of electrons orbiting around nuclei composed of neutrons and protons.

Forces are created by the exchange of discrete packets of energy, or quanta. Different forces are caused by the exchange of different quanta.

In the perspective of the FLPM/IN concept, these forces play crucial roles, yet to be determined, in the internal nucleosynthesis processes within all nuclear energy masses. The FLPM indicates some sort of connection, yet to be understood, between the forces of the quantum mechanics within the masses comprising our SS and its beautiful Phi geometry. Perhaps some day that connection will be deciphered. A quantum leap seems to have been made in that direction in the following section on the string theory.

SUPERSTRINGS AND GEOMETRY

Edward Witten, of the Institute for Advanced Study in Princeton, New Jersey, is the dominant force in the world of theoretical physics. Witten believes that "physics is about concepts, wanting to understand the concepts, the principles by which the world works."

A paragraph from Michio Kaku's book, *HYPERSPACE* states: "Witten's next project is the most ambitious and daring of his career. A new theory called superstring theory has created a sensation in the world of physics, claiming to be the theory that can unite Einstein's theory of gravity with the quantum theory. Witten is not content, however, with the way superstring theory is currently formulated. He has set for himself the problem of finding the origin of superstrings, which may prove to be a decisive development toward explaining the very instant of Creation. The key aspect of this theory, the factor that gives it its power as well as uniqueness, is its unusual geometry; strings can vibrate self-consistently only in 10 and 26 dimensions.

Details of the superstrings concept are beyond the intended scope of this book. Suffice it to quote one of its originators, David Gross: "To build matter itself from geometry -- that in a sense is what string theory does. It can be thought of that way, especially in a theory like the heterotic string which is inherently a theory of gravity in which the particles of matter as well as the other forces of nature emerge in the same way that gravity emerges from geometry."

The most remarkable feature of string theory is that Einstein's theory of gravity is automatically contained in it. According to Witten: "String theory is extremely attractive because gravity is forced upon us. All known consistent string theories include gravity, so while gravity is impossible in quantum field theory as we have known it, it's obligatory in string theory." He adds, "all the really great ideas in physics" are "spinoffs" of the superstring theory.

Quoting again from *HYPERSPACE*: "In conclusion, the symetries that we see around us, from rainbows to blossoming flowers to crystals, may ultimately be viewed as manifestations of fragments of the original ten-dimensional theory. Reimann and Einstein had hoped to find a geometric understanding of why forces can determine the motion and nature of matter. But they were missing a key ingredient in showing the relationship. ...This missing link is most likely the superstring theory. With the ten-dimensional string theory, we see that the geometry of the string may ultimately be responsible for both the forces and the structure of matter."

Surely the beautiful Phi geometry of the spacing of planets in our SS, like the symetries of rainbows, blossoming flowers, crystals, etc., offers vital clues to the relationship that exists among the geometry and the forces and structure of matter. How the geometric forces of the SS relate to the geometric forces and structure of the atom and to all matter created therefrom seems crucial to understanding the true nature of our origins.

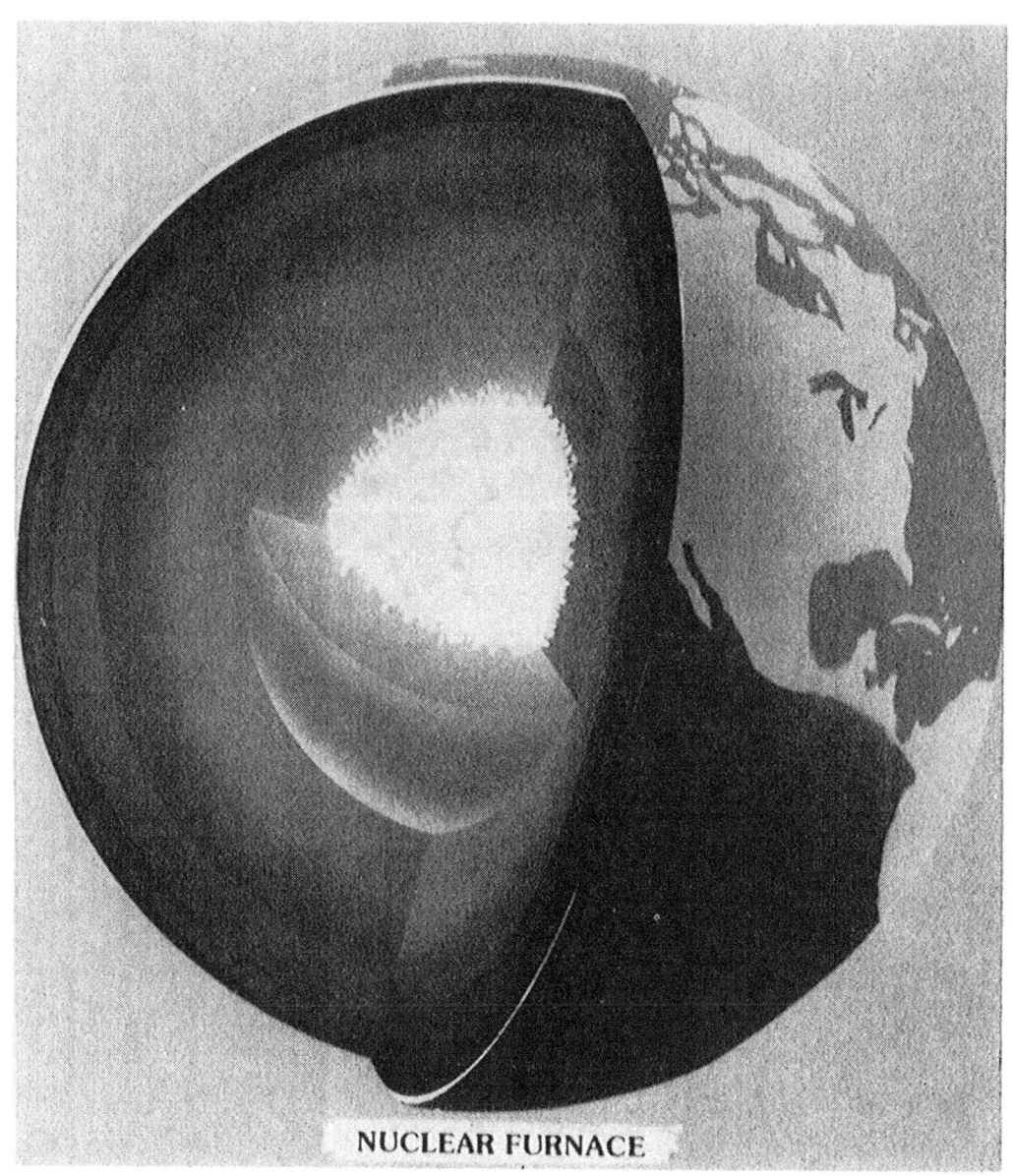

NUCLEAR FURNACE

Earth's nuclear core: A hot, searing, violent, mysterious force that both builds and destroys. Why does mankind seem incapable of grasping its significance in Nature's scheme of creation, of realizing its existence and scope, its power -- a power far exceeding that of billions of hydrogen bombs?

Figure 1

25

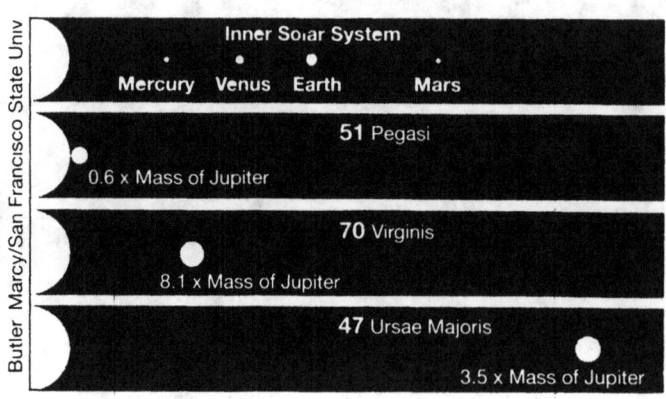

The relative locations of planets in our Solar System and the newly discovered planets orbiting 51 Pegasi, 70 Virginis, and 47 Ursae Majoris.

FIVE STAGES OF PLANETARY EVOLUTION.

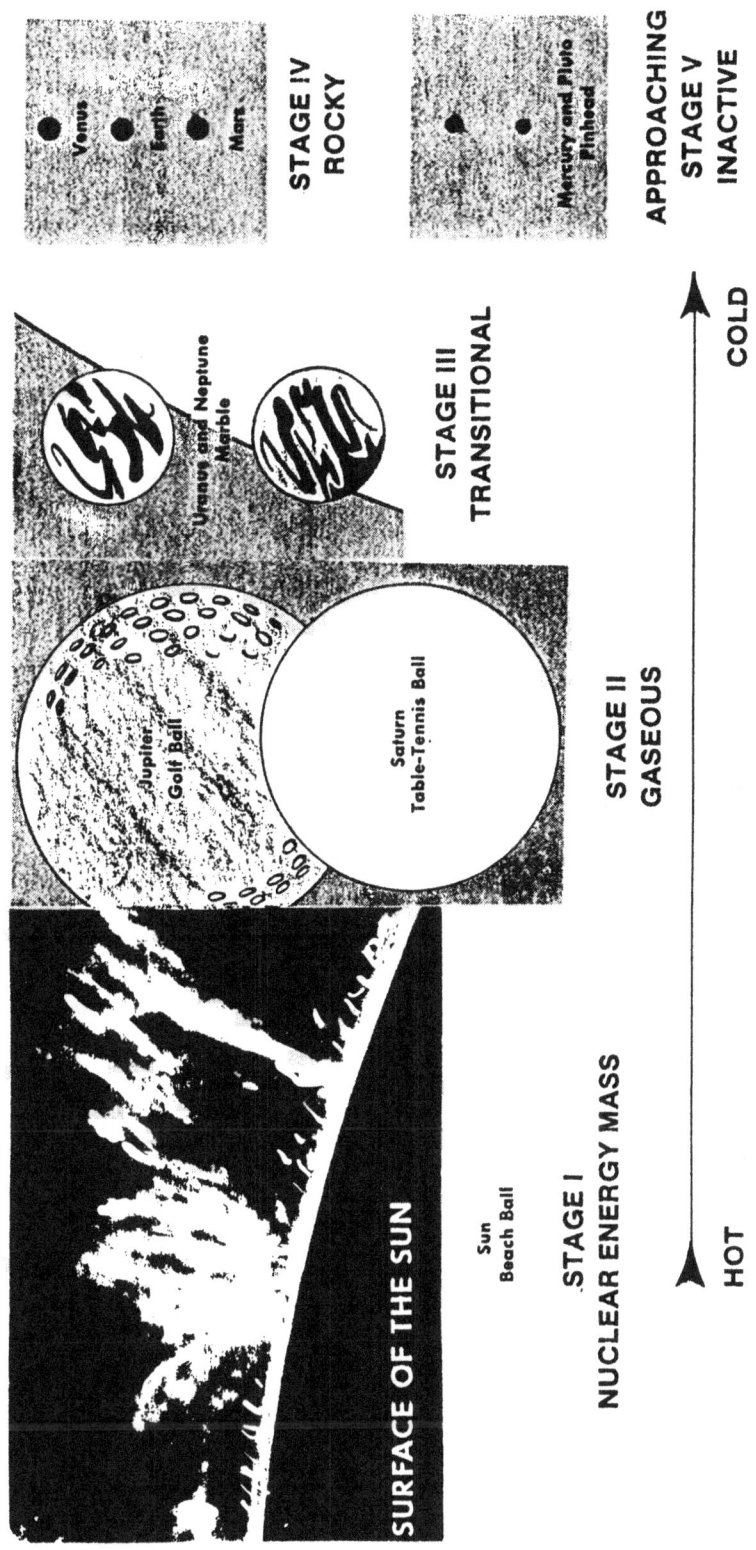

STAGE I
NUCLEAR ENERGY MASS

SURFACE OF THE SUN

Sun
Beach Ball

STAGE II
GASEOUS

Jupiter
Golf Ball

Saturn
Table-Tennis Ball

STAGE III
TRANSITIONAL

Uranus and Neptune
Marble

STAGE IV
ROCKY

Venus

Earth

Mars

APPROACHING
STAGE V
INACTIVE

Mercury and Pluto
Pinhead

HOT ◄————► COLD

HOW PLANETS EVOLVE THROUGH FIVE COMMON STAGES.

The relationship between the size and rate of evolution from energy masses to gaseous planets (Jupiter, Saturn) to transitional combinations (Uranus, Neptune) to rocky planets (Earth, Venus, Mars) and finally to the last stage of inactive (now nearly inactive) crustal spheres (Mercury and Pluto) — (Can include our Moon.).

A. All planetary bodies evolve through these five stages, IAW the laws of thermodynamics and all natural laws.

B. In any solar system, the smaller the sphere, the more rapid its evolution through these stages.

Figure 2

27

We must forsake such beliefs that Earth is flat, that its core is iron or rock, that fuels were made from fossils, that energy is finite, that planets are created from dust and gases, and that continents still drift. Such concepts have too long misled scientists down blind alleys.

Author (1980)

Birth of the Solar System. This painting, sketched by the author and painted by Charles Warner, shows the masses of Jupiter and its moons breaking off from the mother mass, which continues on its path to form Saturn, Uranus, Neptune and Pluto.

Figure 3

Beauty is a conspicuous element in the abstract completeness aimed at in the higher mathematics: it is the goal of physics as it seeks to construe the order of the universe; it ought at least to be the inspiration of all study of life...

John Oman

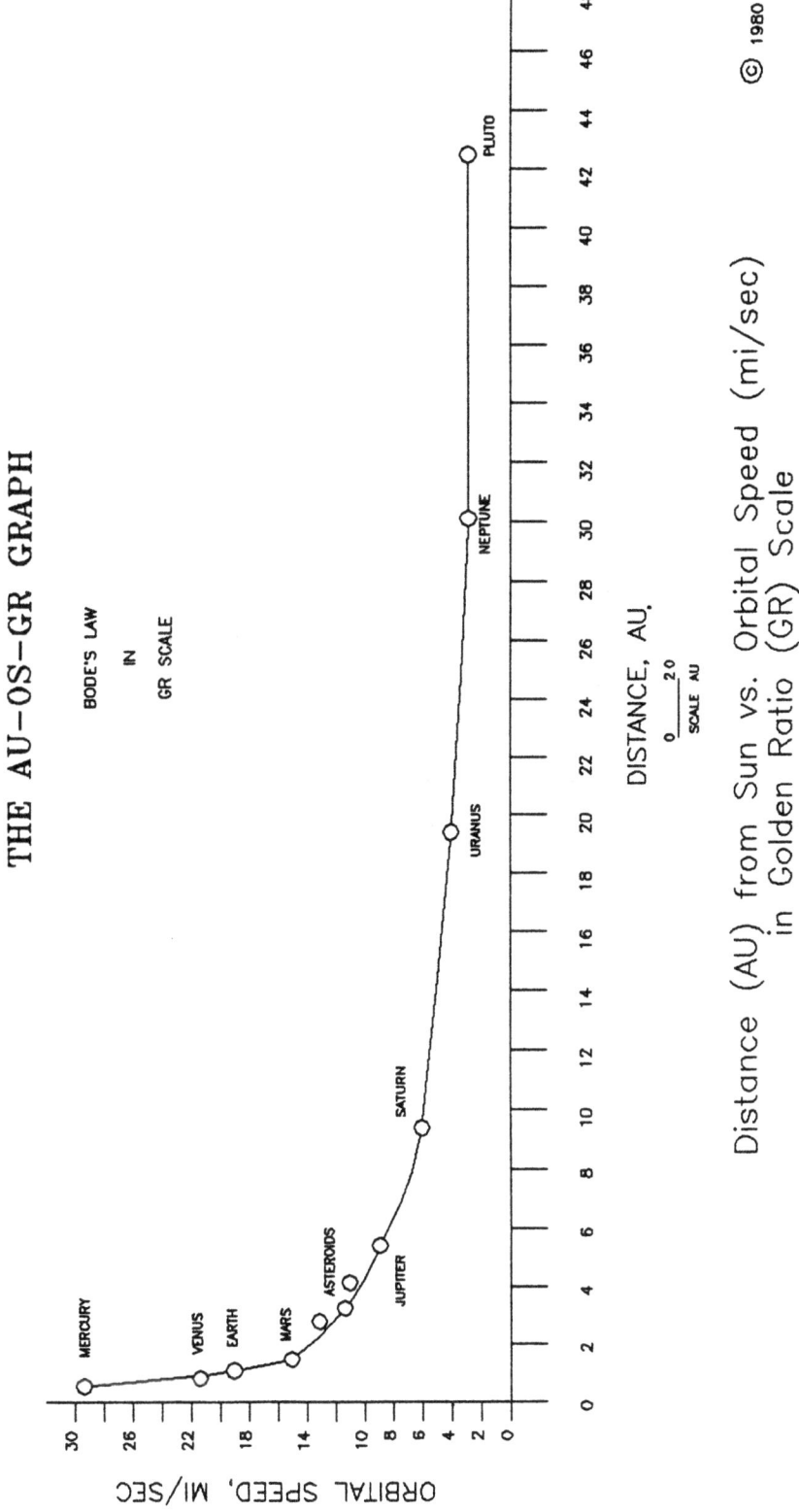

Distance (AU) from Sun vs. Orbital Speed (mi/sec)
in Golden Ratio (GR) Scale

Figure 4

31

Geometry has two great treasures: one is the theorem of Pythagoras; the other, the division of a line into extreme and mean ratio (the Golden Ratio). The first we may compare to a measure of gold; the second we may name a precious jewel.

Johannes Kepler

CHAOS IN THE PLANETARY ORBITS OF THE SOLAR SYSTEM

Deviations of Planets from Original GR Orbits

GR Law	vs.	Bode's Law
AU= 1.618034 X PO	vs.	AU = 0.4 + 0.1X

(Revealing the relationship of Bode's Law to Nature's GR Law governing the geometric Origin of the Solar System)

Planet	Original	Distance, AU Current	Diff.
Mercury	0.618	0.387	0.231
Venus	1.000	0.723	0.277
Earth	1.618	1.00	0.618
Mars	2.618	1.52	1.10
Asteroids	4.236	3.23	1.01
Jupiter	6.854	5.39	1.46

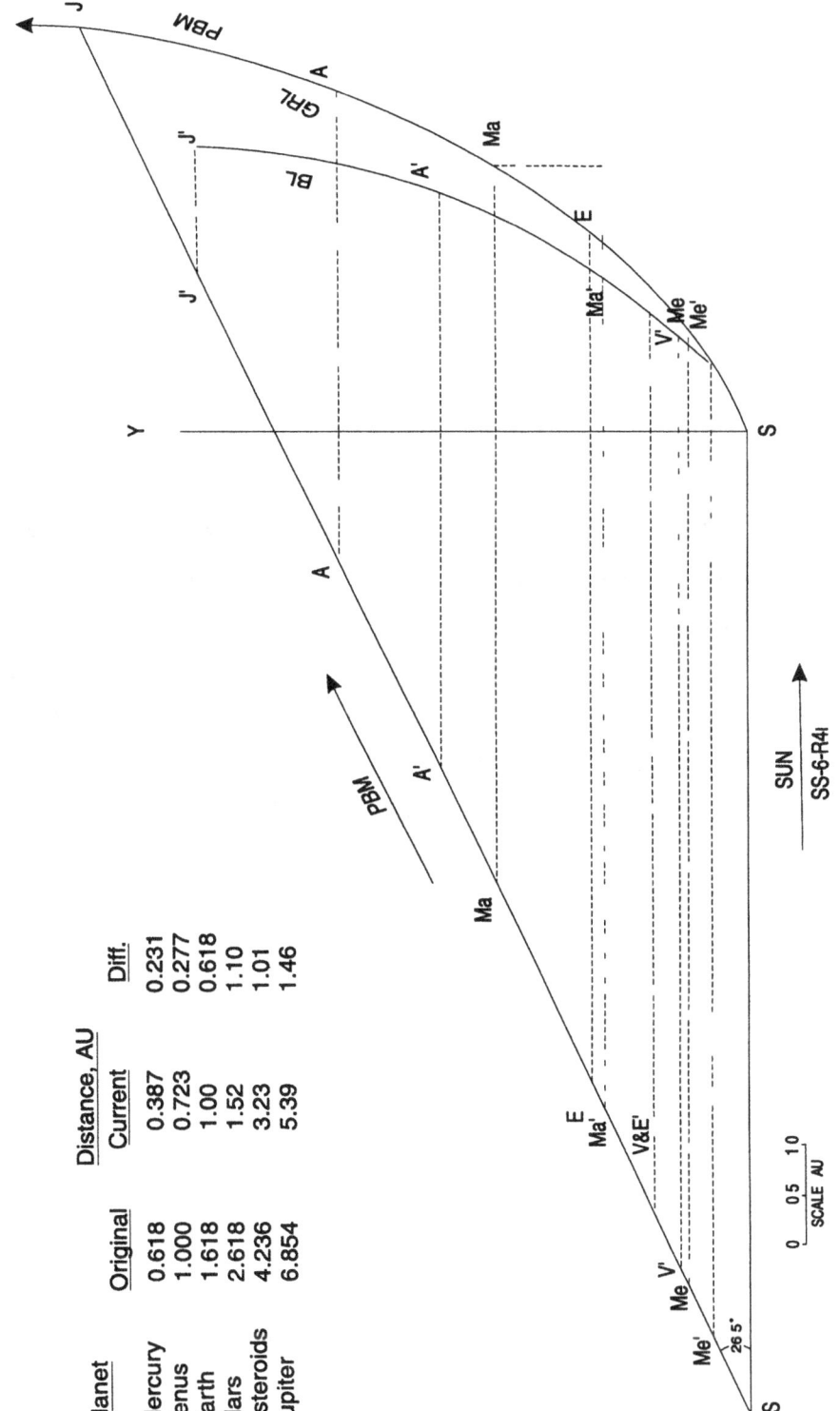

Figure 5

33

On the Golden Ratio:

Nature uses this as one of her most indispensable measuring rods, absolutely reliable, yet never without variety, producing perfect stability of purpose without the slightest risk of monotony... We shall find it flung broadcast throughout all Nature.

C. Arthur Coan

CHAOS IN THE PLANETARY ORBITS OF THE SOLAR SYSTEM

Deviations of Planets from Original GR Orbits

$$\frac{GR\ Law}{AU = 1.618034 \times PO} \quad vs. \quad \frac{Bode's\ Law}{AU = 0.4 + 0.1X}$$

(Revealing the relationship of Bode's Law to Nature's GR
Law governing the geometric Origin of the Solar System)

Planet	Original	Distance, AU Current	Diff.
Mercury	0.618	0.387	0.231
Venus	1.000	0.723	0.277
Earth	1.618	1.00	0.618
Mars	2.618	1.52	1.10
Asteroids	4.236	3.23	1.01
Jupiter	6.854	5.39	1.46
Saturn	11.09	9.50	1.59
Uranus	17.94	19.20	-1.26
Neptune	29.03	30.10	-1.07
Pluto	46.98	42.30	4.68
Z (theo)	76.01	NA	NA

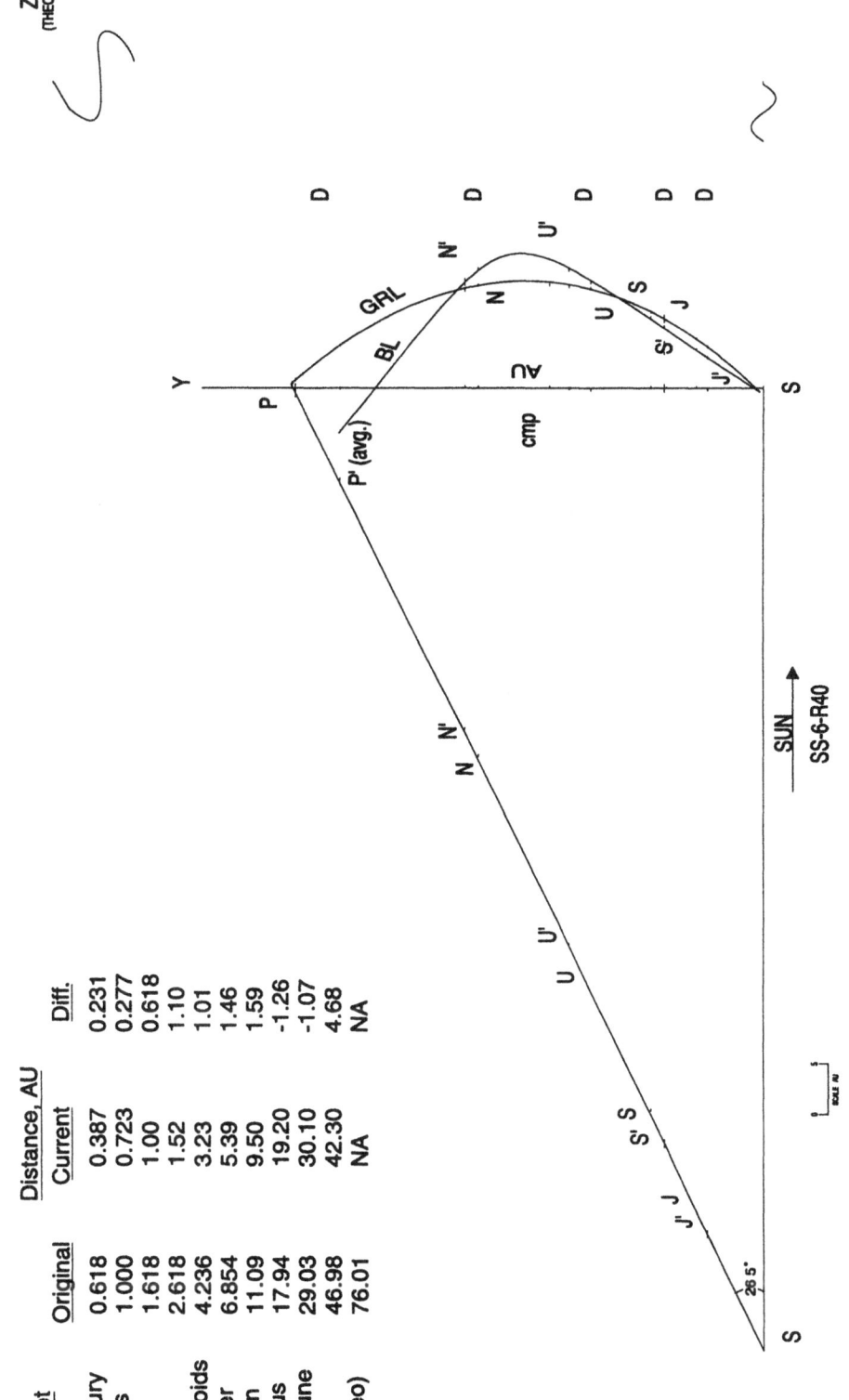

Figure 6

Above all else -- truth.

GEOMETRIC ORIGIN OF THE SOLAR SYSTEM

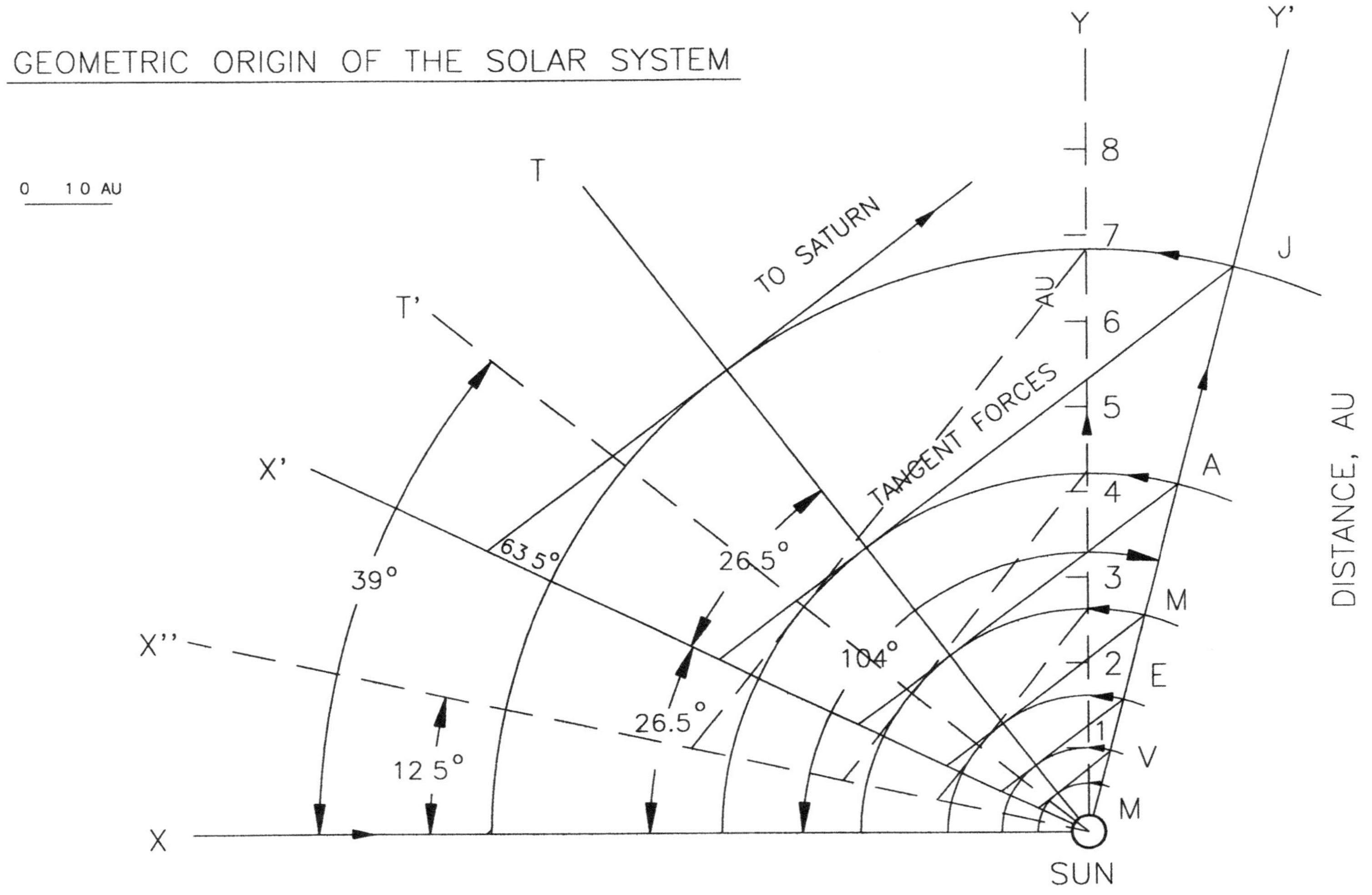

DISTANCE, AU

ILLUSTRATES HOW THE INNER PLANTS WERE PLACED IN ORBITS VIA THE GOLDEN RATIO GEOMETRY OF THE FOURTH LAW OF PLANETARY MOTION.

Figure 7

© 1995

The important thing is not to stop questioning.

Albert Einstein

GEOMETRIC ORIGIN OF THE SOLAR SYSTEM

0   20 AU

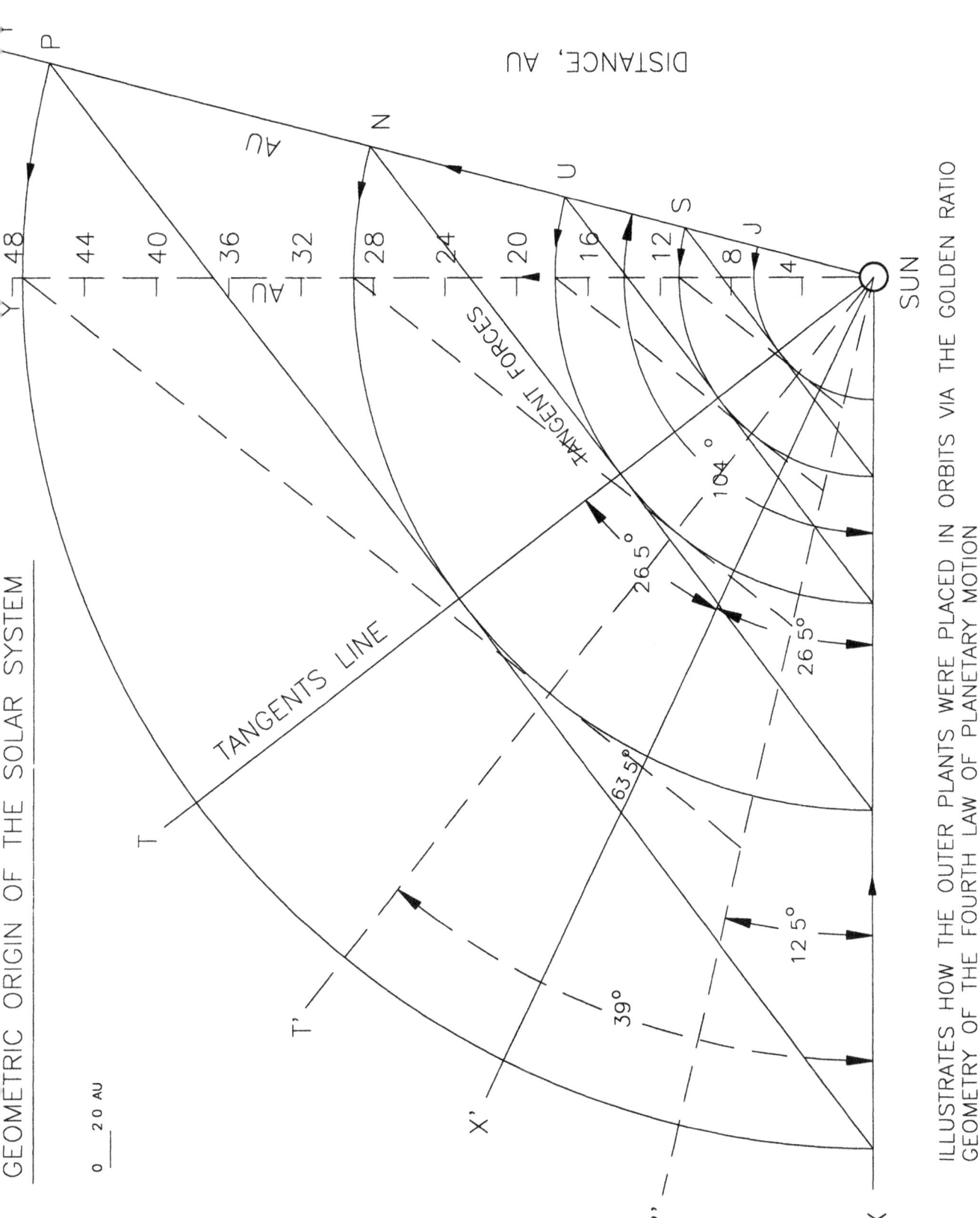

DISTANCE, AU

ILLUSTRATES HOW THE OUTER PLANTS WERE PLACED IN ORBITS VIA THE GOLDEN RATIO GEOMETRY OF THE FOURTH LAW OF PLANETARY MOTION

Figure 8

© 1995

39

Simply because it's impossible does
not mean it can't be done.

THE AU–OV–GR CURVE

FOR
THE GEOMETRIC ORIGIN
OF
THE SOLAR SYSTEM

DISTANCE vs. VELOCITY

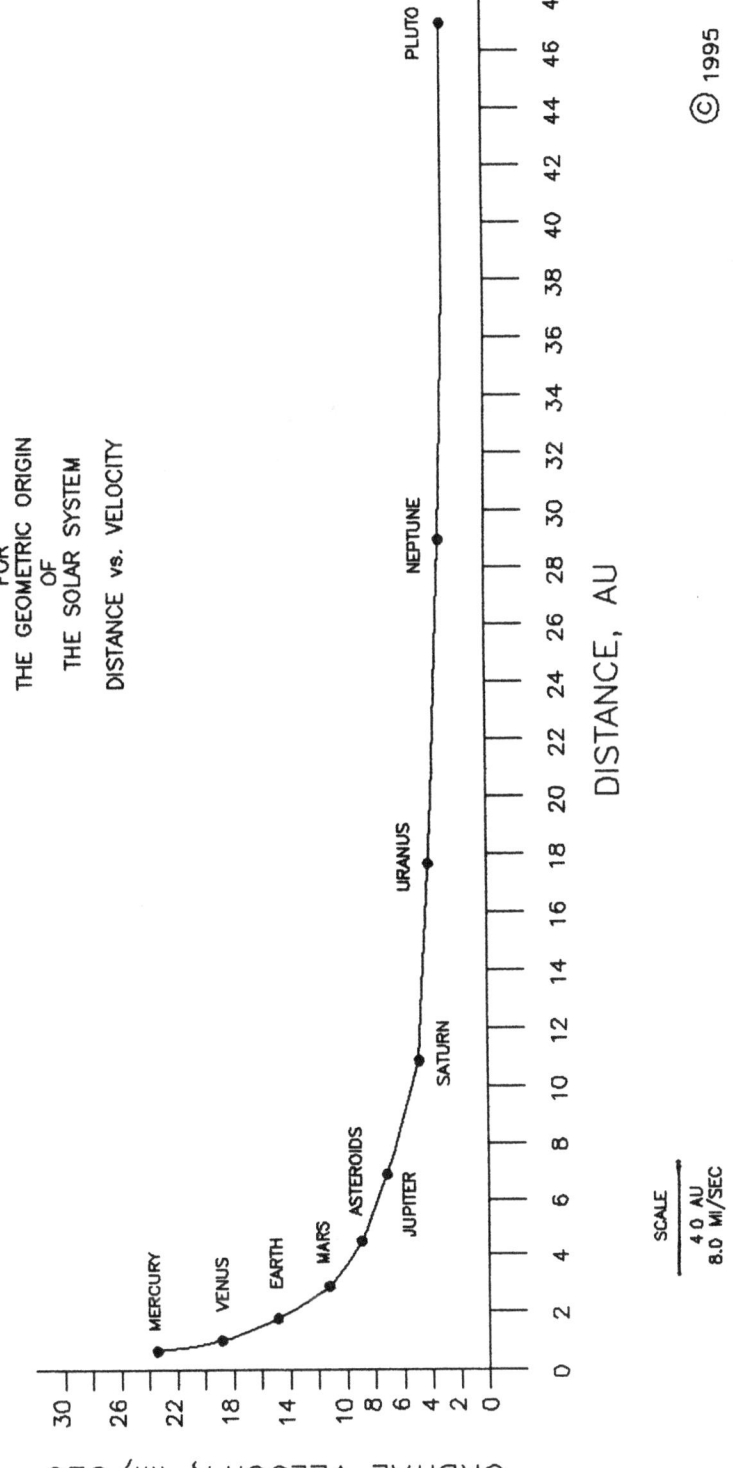

© 1995

Figure 9

41

It is nothing short of a miracle that modern methods of instruction have not yet entirely strangled the holy curiosity of inquiry.

Albert Einstein (1949)

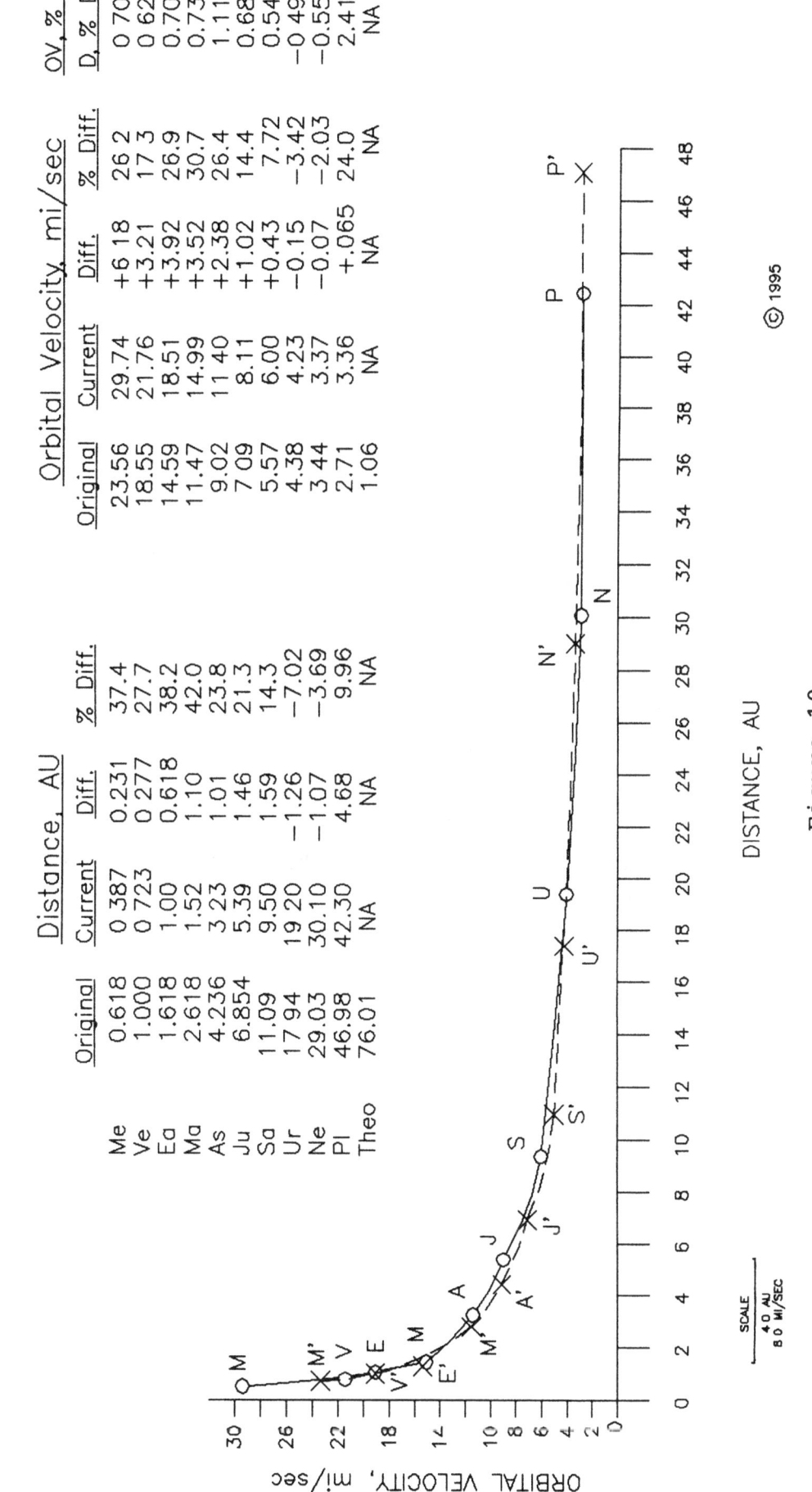

THE AU-OV CURVES (SUPERIMPOSED)

(GR vs. BL)

o CURRENT, BL
X ORIGINAL, GR

DISTANCE AND VELOCITY CHANGES IN THE ORBITS OF PLANETS

	Distance, AU			Orbital Velocity, mi/sec				OV, % Diff.	
	Original	Current	Diff.	% Diff.	Original	Current	Diff.	% Diff.	D, % Diff.
Me	0.618	0.387	0.231	37.4	23.56	29.74	+6.18	26.2	0.70
Ve	1.000	0.723	0.277	27.7	18.55	21.76	+3.21	17.3	0.62
Ea	1.618	1.00	0.618	38.2	14.59	18.51	+3.92	26.9	0.70
Ma	2.618	1.52	1.10	42.0	11.47	14.99	+3.52	30.7	0.73
As	4.236	3.23	1.01	23.8	9.02	11.40	+2.38	26.4	1.11
Ju	6.854	5.39	1.46	21.3	7.09	8.11	+1.02	14.4	0.68
Sa	11.09	9.50	1.59	14.3	5.57	6.00	+0.43	7.72	0.54
Ur	17.94	19.20	−1.26	−7.02	4.38	4.23	−0.15	−3.42	−0.49
Ne	29.03	30.10	−1.07	−3.69	3.44	3.37	−0.07	−2.03	−0.55
Pl	46.98	42.30	4.68	9.96	2.71	3.36	+.065	24.0	2.41
Theo	76.01	NA	NA	NA	1.06	NA	NA	NA	NA

DISTANCE, AU

© 1995

Figure 10

ORBITAL VELOCITY, mi/sec

SCALE
4.0 AU
8.0 MI/SEC

43

We must look to some other cause, which can gradually and permanently change the profiles of land and sea bottom. I hold this cause to be an independent one. I also hold the two processes to be distinct and to have no necessary relation... Whatever the cause of great regional uplifts, it in no manner affects the law of isostasy. The real nature of _that uplifting force_ is a mystery.

Clarence Dutton (1841-1912)

Chapter II

HOW PLANETS EVOLVE

THE FIVE STAGES OF PLANETARY EVOLUTION

All planets began as tiny, fiery stars, each placed in orbit in full accord with the Laws of Planetary Motion. Each is destined to obey these Laws as long as it remains in orbit around its sun.

The transition of our planets from nuclear energy to matter has required some five billion years of evolution to reach the current stages observed in the nine planets and the asteroids of our SS. The ten orbits of planetary masses contain clues to their origins, but remain mysteries wrapped in enigmas. The clues are many and straightforward, poised for proper interpretations.

The Phi geometry of the SS (explained in Chapter I) revealed how and where each planet was placed in orbit, and how much each has been displaced from its original orbit. Knowing this, we can uncover many clues that reveal how each planet evolved by means of natural laws to its present stage of evolution. This is possible within the realm of the natural laws of physics and chemistry, and can be accomplished without speculations or assumptions that too often mislead.

How is it possible for planets and moons to evolve from nuclear fireballs into the spheres of the SS as observed today in their geometrically spaced orbits? Our Sun furnishes significant clues to the evolution of planets by natural laws. An atomic forge, it is readily recognized as one in which fission-fusion reactions release energies in the form of heat and light from atomic nuclei undergoing constant changes. For example, four hydrogen nuclei fuse together to create one nucleus of helium, and almost 1% of the original weight or mass of the hydrogen material is changed into heat and light.

The glowing hot gases of our Sun are made of the lighter elements that comprise the huge gaseous planets and are found in the crust, oceans and atmosphere of Earth. The most common elements in the Sun are hydrogen, helium, calcium, sodium, magnesium and iron; the number of other elements include oxygen, nitrogen and carbon. The main point here is that atoms are created in situ from the energy of nuclear masses under extreme conditions of temperature and pressure. The big question becomes: How does a nuclear mass (a small sun) metamorphose into a planet or moon?

The first phase of transformation begins with the accumulation of countless numbers of atoms, ions and simple compounds that eventually form an enveloping spherical blanket high above their source (the energy mass, now identifiable as the core). Simultaneously, similar matter, as with the Sun, accumulates on the extremely hot "surface" of the core. During billions of years, virgin matter pours from the manufacturing plant below, slowly filling the huge space between mass and the spherical blanket to form clouds of chemical vapors throughout the towering atmosphere. Scientists see the results in Jupiter and Saturn, both in the second stage of planetary evolution, and classify them as gaseous planets.

In the second phase of transition, the voluminous clouds of towering atmospheric matter gradually close in on the hot interface, and precipitation of its chemical vapors is initiated. The precipitates are instantly repelled by the heat; the gigantic battles between the evaporation and the condensation forces rage on for eons. The atmospheric matter is trapped between the unbearable heat of the energy mass and the super coldness of outer space. In time, matter gains a toehold in the form of the first tenuous liquid layers, eventually followed by thin, tenuous crustal formations. Uranus and Neptune are into this third stage of evolution, although their outward appearances may or may not justify their classification as gaseous planets.

During the third phase in the evolution of a planetary sphere, shallow seas cover the entire surface, cooling it sufficiently to permit formation of a more permanent crust. The crust thickens as a function of time, a continual process for billions of years. Newly created matter from volcanic outpourings of lava, elements, water, gases, and compounds of great varieties continually build and alter atmospheres and land systems.

With these dramatic, often violent, and persistent environmental changes, species, where feasible have their beginnings; they come, they flourish, and they vanish into extinction as the ever-changing environment dictates. They adapt or they perish. The era of the dinosaurs is a prime example -- perhaps the most popular of all time. Their history offers powerful evidence that Earth is a self-sustaining entity creating its own systems from within, always maintaining control of the creation and extinction of its species. Other prime examples include the life that exists around the "black smoker" vents discovered on deep ocean floors, along with outpourings of virgin materials that never cease. Even more voluminous outpourings of virgin materials occur from the baseball-like seam circling the globe, as often viewed on TV screens.

There is no need to look to outer space for answers to the origin and extinction of species. Earth's multi-layer crust offer clues to events effected by virgin ejecta, including the iridium layer at the K-T boundary, as it continually thickens over time. Although other layers of iridium have been found in other eras, this particular iridium layer has been interpreted erroneously as coming from outer space and wiping out the dinosaurs. In reality, it is a layer of ejecta matter created within Earth via nucleosynthesis under extreme, specific conditions of temperature and pressure -- as were the other iridium layers.

A similar example of a puzzling heavy metal layer was discovered recently on Venus (*Science* 5 Jan 1996). The loftiest parts of the planet, like the highest peaks on Earth, are covered by a perpetual frost -- a coating of the lustrous, silvery-white element tellurium. This explains why the highlands of Venus appear so bright in radar images of the surface. Telllurium has just the right electromagnetic properties to explain the radar brightness of Venus, and just the right melting point to coat its highlands, but not its plains.

Planetary scientists now argue that the metal, nearly as rare as gold, could be spewing from volcanoes as metallic vapors that would freeze out onto the cooler highlands. This is in full agreement with the 1973 idea that has developed into the FLPM/IN concept during the past 22 years, and is powerful confirmatory evidence that the iridium layers on Earth were made in the same manner via the nucleosynthesis that exists in both (and all) planets. Also, it is powerful evidence of the ever-changing planetary environments that spawn and eventually kill the species that cannot adapt to these changes: the true cause of the demise of the dinosaurs.

As planets evolve from tiny, fiery stars into crusty old spheres with ever-diminishing nuclear cores, startling things happen. In Earth's case, the hot, humid environment spawned the first small wonders of life. Bacterial life spawned in temperatures of a few hundred degrees before the surface cooled sufficiently to allow the formation of eubaryotic cells. It was only a matter of time and additional cooling before various forms of life could evolve and be sustained.

DINOSAURS: THE REASONS FOR THEIR EXTINCTION

Eventually, Earth's average temperature cooled to the mid-100s, with high humidities, heavy rains and vicious thunderstorms prevailing: ideal conditions for plush tropical forests worldwide. The steamy forests were fully capable of sustaining huge life forms -- a real paradise for giant animal life. Dinosaurs thrived on the lush plant life and smaller jungle animals during the 140 million years of the Mesozoic Era, often referred to as the Age of the Dinosaurs.

During this rich time span, environmental conditions were changing ever so imperceptibly. Temperatures gradually declined as Earth's crust thickened, its insulation propensities ever increasing to block out the heat emanating from within. The food supplies of Nature's lush forest began to diminish gradually, while the high oxygen content of the atmosphere declined proportionally over the ages.

Vicious, fiery eruptions from Earth's hot interior were commonplace -- the order of the day -- as evidenced by the multi-layers of ejected materials that together with sedimentary layers, comprise the crust. With each passing mill- ennium, the composition of the erupting materials changed in accordance with the prevailing internal conditions of both temperature and pressure as functions of time that determined the type of end-products added as new crust. Each new layer of material contributed its effects, good or bad, to the changing surface and atmosphere. These crustal layers now serve as clocks and books from which mankind extracts their recorded history.

With the passing of time, the dinosaurs adapted as best as they could, while slowly declining in numbers as the environment became ever more hostile. Gradually, Nature began delivering its coup de grâce from the bowels of the mother Earth that had spawned and nurtured the giants through almost 140 million years, and now was dictating their rapid demise, relatively speaking. Internal conditions combined to emit a series of violent eruptions, spewing the iridium that now is found worldwide in a well-defined layer containing concen- trations from 10 to 100 times the normal levels of this rare element.

Analyzed sediments from this K-T boundary layer that marks the end of the dinosaur era revealed that the iridium enrichment, along with other chemical anomalies found there, were deposited over a period of 10,000 to 100,000 years or more.

These anomalies are more consistent with volcanic rather than meteoritic origins. The fact adds significant confirmatory evidence to the IN facet of the FLPM/IN concept pertaining to eruptions of virgin materials that persistently change the surface of Earth, and thereby control the destinies of all its creatures. In this case, the dinosaurs and 90 percent of all genera of protozoans and algae, along with 60 to 75 percent of all species, disappeared from Earth because of the changes wrought via nucleosynthesis within the planet.

During the past few years, a debate has persisted over the source of the iridium in the K-T boundary layer. Some scientists believe that the element came from outer space, brought in by a crashing asteroid that cloaked Earth with a cloud of dust, resulting in darkness, suppression of photosynthesis, the collapse of food chains and ultimately, mass extinction. However, such happenings do not require anything from outer space; Earth is capable of stirring up its own mess.

The K-T boundary at Gubbio, Italy was re-sampled in 1990 in detail for both iridium content and magnetostratigraphy by a team that included both the terrestrial and impact proponents. The results confirmed that the iridium anomaly covers about three meters of the vertical section, representing about 500,000 years of deposits. There now seems little doubt that the iridium was not deposited there by the impact of a large asteroid. The evidence clearly shows that the deposits are the result of volcanic actions, thereby adding more powerful support for the FLPM/IN concept.

Many scientists still attribute this demise to an extraterrestrial crash of a huge meteorite or comet. But the evidence against this belief grows stronger. Recently, Nicola Swinburne and her co-workers found glass spherules and high concentrations of iridium within 61-million-year-old rocks in West Greenland. When researchers detect such evidence in rocks of K-T boundary age, they often misinterpret it as a sign of an impact.

These materials were found in volcanic rock, increasing the probability that an eruption created the spherules and the iridium layer. The Greenland rocks also contain large chunks of nickel-iron metal, a principal component of some meteorites. These findings fit precisely into the FLPM/IN concept in which all of these materials are simply ejecta created via nucleosynthesis within Earth's nuclear core.

Ice core drillers at Russia's Vostok Station, atop the great ice sheet of East Antarctica, recently passed 3000 meters -- a depth at which the ice is about 300,000 years old. Analyses of air bubbles trapped in the ice have confirmed that levels of carbon dioxide and methane were higher between glacial periods than during them. These findings illustrate how our atmosphere changes over time as a result of subtle changes in Earth's internal conditions that determine the types and quantities of elements (matter) created at any given time within the nuclear core. *The creation and expelling of significantly less carbon dioxide and methane to the atmosphere resulted in eras of drastic cooling, now identified as glacial periods (just the opposite of the current warming period).*

The most obvious and critical factor determining the rate of planetary evolution is size. The smaller the planet, the more advanced its stage of evolution. Scientists observe Earth, Venus and Mars, and classify them as rocky planets (the fourth stage of planetary evolution).

In the fifth and final stage, planets and moons become inactive spheres. Core energies become depleted to the point that no new material can be created within. Outpourings and seismic activity cease. Electromagnetism may or may not be detectable. Mercury, Mars, Pluto and Moon are examples of smaller, nearly inactive spheres. Each has only traces, if any, of seismic activity and electromagnetism remaining.

For example, Moon's very faint seismic activity is revealed in tiny moonquakes, indicating that its nearly depleted core is still capable of creating the weak out-gassings observed in craters. Moon once generated its own magnetic field, which may have been nearly twice as strong as the present-day magnetic field of Earth, according to S.K. Runcorn and colleagues, who used <u>magnetized lava rock</u> from Moon as evidence. This verifies that <u>our</u> Moon's <u>small nuclear core originally was larger than Earth's energy core is today</u> (and both are being depleted).

The proof of the declining strength of Moon's magnetic field is a strong indication that such magnetism is of nuclear energy origin rather than of a steady-state iron core, rock or other origin. This large decline is attributable to the dwindling size of the core as its energy transforms into matter. An iron core, or any other type core, would not dwindle.

From this discussion of the transition phases of planetary evolution, two conclusions can be drawn:

1. All planets and spherical moons evolve through five common stages of evolution via nucleosynthesis by means of natural laws.

2. In any SS, the smaller the mass, the more rapidly it evolves through the five hot-to-cold stages.

One last comment on current beliefs about the origin of our Moon (the giant impact scenario and its most plausible alternative, coaccretion) seems appropriate: In the words of one scientist, "they are more a testament of our ignorance than a statement of scientific knowledge."

FOUR ORIGINAL CLUES TO PLANETARY NUCLEAR CORES

A number of observations and facts contributed to the conclusion in 1973 that a nuclear mass, rather than rock (then the prevalent theory) or iron (the current prevalent belief), exists as Earth's core. First, the three-layered system of energy fuels throughout the crust made it obvious that a central source of nuclear energy was essential for creation of elements and compounds comprising these fuels: gas, petroleum, and coal. The predicted vast stores of deep gas, composed of tremendous volumes of the two elements, carbon and hydrogen, strongly indicated only one possible source for these building blocks comprising methane gas -- the starting point for creation of hydrocarbon fuels.

The second and most common observations are the tremendous out-pourings of lavas, gases and other matter from volcanoes and rifts throughout Earth's history. Examples include the lava flows that played major roles in building the continents of North America, Europe and India (Deccan Traps), the mile-high multi-layer Grand Canyon and numerous other layered systems throughout the crust.

The third clue is the radioactivity within Earth's crust -- known to be a natural byproduct of fission-fusion reactions in nuclear masses. This obvious interpretation of this powerful evidence remains in direct opposition to the persistent belief that credits radioactivity as the source of internal heat, rather than the byproduct of the interior reactions. However, the fact that fission-fusion reactions inside nuclear masses (e.g.; nuclear bombs) do produce radioactivity (as well as heavy elements 99 and 100) positions this as the most logical reason for its presence in the crust. One must question why the radioactivity found in crustal matter is always cold. And how would radioactivity get inside Earth without the presence of a bona fide source to produce it?

The fourth original clue indicative of nuclear cores inside planets and moons can be observed daily: the high mountains and uplifted plateaus found on Earth, Mars, Moon, and more recently, as predicted by this concept, Venus. The presence of these similar characteristics on these spheres is strongly indicative of a common cause: the creation and uplifting of their surface features by very powerful forces within each one. Only nuclear power, in conjunction with isostacy, seemed capable of wielding such magic. These observations and interpretations made it easy to predict that other planets and moons have undergone similar processes. Time and space probes have proven the accuracy of this prediction.

RECENT CLUES TO NUCLEAR CORES IN PLANETARY SPHERES

A number of clues indicative of nuclear cores have been discovered during the last few years. Foremost among the predicted evidence is the discovery in 1987 that Earth's center is hotter than the surface of the Sun. The inner core has a temperature of about 12,420°F, scientists from the University of California at Berkeley and the California Institute of Technology reported in the April journal, *Science*. They calculated temperatures of 11,900°F for the boundary between the inner and outer cores, and 8,640°F for the outer core-mantle boundary. In comparison, the surface temperature of the Sun is about 10,000°F.

The researchers based their experiments on the assumption of an iron core at Earth's center and a pressure of 49 million pounds per square inch. Their finding was surprising to them because it suggests that <u>the core</u>, not the enveloping mantle, <u>is</u> <u>the</u> <u>source</u> <u>of</u> <u>much</u> [actually all] <u>of the</u> <u>internal</u> <u>heat</u>.

"Thus, the forces that drive the plates and give rise to earthquakes and volcanoes have their origins in the Earth's core," said Thomas Ahrens. "This provides us strong insight into how Earth works."

While this insight is a giant leap forward from the rocky core concept of a few years ago, it falls far short of the true situation inside Earth: a nuclear core

52

with temperatures ranging into the millions of degrees -- perhaps into the 100 million degrees range believed to be a requisite for creating uranium, Earth's heaviest stable element.

The core temperatures calculated by Ahrens *et al.* are far too high to have been generated by the slow decay of radioactive elements. Further, such calculations are invalidated by the false assumptions of an iron core and 49 million pounds pressure, which result in temperatures far below reality.

Viewing these factors in the light of the second law of thermodynamics, one must conclude that the extreme heat was present from the beginning over 4.6 billion years ago, even before Earth began forming its crust. This evidence adds dramatically to the FLPM/IN concept of an original nuclear mass transforming into matter that formed our atmosphere and crust.

The second important piece of new evidence supporting this concept is the discovery made in the 1980s by seismic tomographers exploring the interior of our planet via sound waves. By slicing open Earth with tomographic techniques and modern computers, geophysicists have uncovered features from the crust down to the core. Topographic maps prepared by scientists at Caltech and Harvard show 'blobs' of hotter material rising from the core-mantle boundary, while cooler masses are sinking from the upper mantle into the interior.

Contradictory to the serene, onion-layer concept, the outer core does not have a smooth, bland surface; rather it consists of alternating deep valleys and mountains. The scale of these deep depressions and elevations is estimated to be between five and ten kilometers -- greater than Mount Everest -- in some locations. These findings remind one of Sun flares, conforming to the visualized violent image of a nuclear mass encapsulated in the mantle of hot, molten matter of its own making.

The third piece of predicted and confirmatory evidence favoring the new concept came with the discovery in the 1980s of the simultaneous increases in sea levels and polar ice sheets. According to a 1985 National Academy of Science report, sea level is rising about one-tenth inch annually, but scientists don't know why. They have been unable to put the responsibility on the greenhouse effect or on the polar ice sheets.

However, when viewed in the perspective of the FLPM/IN concept, the answer becomes obvious: the internal creation of virgin water, as with observable land increases, is an ongoing, never-ending process.

The best examples of virgin water can be seen in the outpourings from deep-sea hydrothermal vents. The Juan de Fuca ridge in the Pacific Ocean has hydrothermal vents spewing jets of mineral-laden water at 662°F in a fairly continuous stream. The minerals are making their surface debut as new matter from Earth's transformation factory.

Scientists believe that megaplumes (large columns of warm water) come from fields of these vents. But the megaplumes represent an explosion of fluids, like a giant underwater burp. Multiplied many times throughout the world, the result is a steady increase in sea levels. For virgin water and other new matter, there are, of course, many other outlets: volcanoes, rifts, etc.

In September, 1981, government scientists began analyzing metal-bearing chunks spewed out of an undersea volcano 270 miles west of Oregon. H.E. Clifton, Chief of the USGS's Pacific-Arctic Branch of Marine Geology, reported that "new earth crust is actually being made" by the volcano.

Other strong arguments for creation of virgin materials that form Earth's crust include the "black smokers" discovered in 1979 by a team of American, French, and Mexican investigators. They observed turbulent black clouds of fluid billowing up from chimney-like vents, much like factory smokestacks (which, in reality, they are). The venting fluid, a metal-rich hydrothermal solution was measured at 350°C. Mixing with the ambient sea water causes copper-iron-zinc sulfides to precipitate as fine black particles suspended in the plumes.

The grade of the metals is comparable to that of many ancient massive piles of sulfides on land: 31% zinc, 14% iron, 1% copper, plus small amounts of silver and gold. On the island of Cyprus some 90 large deposits of copper-iron-zinc sulfides occur as saucer-shaped bodies up to hundreds of meters in diameter. They fill depressions in volcanic lavas that erupted on the sea floor some 85 million years ago. Still further back in geological time, similar hydrothermal convection systems were active 2.7 billion years ago in rocks of the Archean period, now exposed in the eastern Canadian shield.

Minerals and lavas, water and land, air, gases, oils, sulfides and other chemicals, etc. constantly pouring from an internal source over eons of time. Such observations continue to add credibility to the theory of Internal Formation of Fuels and Elements (IFFE) of 1973, later changed to the Internal Transition of Energy to Matter (I-T-E-M) theory, and now identified by more recent terminology as the Internal Nucleosynthesis (IN) concept. In combination with the new Fourth Law of Planetary Motion, it has evolved into the all-encompassing FLPM/IN concept that explains our planetary origins and evolution from the beginning to the present day.

The reason for these continuous outpourings is a crucial point in the concept. The transformation of energy into molecules of matter entails a tremendous expansion, relatively speaking. In such transformation, each molecule expands into far more space than had been occupied by the energy from which it was made. This forced expansion of uncountable molecules is what creates ever more pressure and, consequently, ever greater temperature within the nuclear energy core. These dramatic increases in the demand for space and in temperature account for the capability of Earth to produce all heavy elements up to uranium.

Simultaneously, the expansions result in a steady, but imperceptible increase in the size of Earth. As it expands, the crust cracks, giving the surface an appearance similar to the cracked shell of a hard-boiled egg.

The fourth piece of predicted and confirmatory evidence for the IN concept can be illustrated by events occurring in Lake Nyos in Cameroon, West Africa. One night in 1986, Nyos experienced a large burst of carbon dioxide from the lake depths. The spreading gas snuffed out the lives of 1700 people. Two years earlier the same type gas had burst from Lake Monoun, killing 17 people. The other 37 lakes in Cameroon posed no immediate danger, although they, too, are nestled in volcanic craters. Scientists remain puzzled by the processes involved and the true source of the gas.

Nothing but a warm, mineral-laden subsurface spring seemed capable of delivering the type of ions identified as increasing quantitatively. The composition and warm temperature of the bottom water point to a hot, deep spring feeding it from below. Much like deep-sea hydrothermal vents and our familiar volcanoes, the volcanic craters holding the lakes simply serve as outlets for the virgin gases. Worldwide and on other planets and moons many similar outlets exist for such out-gassings.

Methane and hydrogen sulfide are examples of gases discovered on a number of SS bodies besides Earth. Hydrogen sulfide, found for the first time (1989) outside our planet, is present on the surface and in the atmosphere of Io, one of Jupiter's moons. More recently and seemingly more strange, French astronomers reported the presence of hydrogen sulfide in comets Austin and Levy. These crucial discoveries confirm that gases are consistently present in all SS bodies suspected of being powered by nuclear interiors. Comets, as we shall see later, are no exception.

A good example to consider here is the gaseous atmosphere of our nearest neighbor, Venus. Known for some time, its bright envelope is 96% carbon dioxide with a substantial admixture of argon. Such an atmosphere argues strongly for out-gassing from a hot interior capable of selectively creating such voluminous gases. Further, recent mappings of the surface of Venus reveal that its unique surface characteristics, conforming to predictions, are readily explainable by principles of the FLPM/IN concept discussed previously. Its excessive atmospheric heat contributes to its uniquely flowing surface features.

WHY ELECTROMAGNETIC FIELD STRENGTHS VARY

One of the most puzzling anomalies in science concerns the origins of the electromagnetic fields embodied in the spheres of our SS. Why do they exist and why do their strengths range so widely between very weak and very strong from sphere to sphere? Do they remain constant for billions of years?

A comparison of the field strengths of Venus, Earth and Mars should offer some insight into the reason for the wide range of electromagnetic strength from sphere to sphere. Venus is almost as large as Earth, but Mars is only one-seventh as large. The fields of Venus, a very slowly rotating planet, and of Mars, a very small planet, are much weaker than Earth's. At this writing, NASA scientists have concluded that if Mars does have an intrinsic magnetic field, it is not of any consequence. Likewise, although nearly the size of Earth, Venus has been identified as a planet with little, if any, electromagnetism. Additionally, our Moon, being small and having only one rotation for each revolution around Earth, shows no detectable magnetic field.

Further, the highly tilted and offset magnetic field of Uranus, a midsize sphere, is the strangest one among planets. The offset from the planet's center causes its magnetic field strengths at the surface to vary by a factor of ten between the north and south magnetic poles. Adding to the evidence, the two largest planets, Jupiter and Saturn, both with faster rotational speeds than Earth, can be expected to have much stronger EM fields. And they do. Mercury, the smallest planet, has a field strength of only one percent of Earth's, and a thin atmosphere. Both clues are indicative of the presence of a small, active remnant of a nuclear core.

There are two basic principles of electromagnetism: Electricity in motion produces a magnetic field, and a magnetic field in motion across an electrical field produces an electromotive force. Combining these principles with the above observations, one can make a general rule: The electromagnetic field strength of each SS sphere is a function of its core size, speed of rotation, (and perhaps its velocity of revolution) and the angle of inclination of the axis.

The electromagnetic field strength (EFS) rule can be true only in the concept of nuclear cores. Thus, a fifth new confirmation of this revolutionary IN idea appears to have been established. If successful, it will further verify the validity of the total FLPM/IN concept.

In contrast to the new concept, prevailing beliefs dictate the existence of various types of cores in the individual spheres, and serve as generators of their magnetic fields. For example, Earth's core consists of iron, either in the molten state or a crystallized solid. Using Earth as an example, one of the main problems with such non-nuclear cores can be seen in a comparison of potential strengths between a field created by an iron core versus a field created by a nuclear energy core; e.g., the Sun. The powerful, extensive magnetic field created by the Sun's nuclear mass reaches many billions of miles to the edge of the SS -- billions of times greater than the size of the Sun.

Experiments with iron magnets and electric generators quickly reveal their very limited ranges: only relatively small multiple of their sizes. In this perspective, the choice of a nuclear core rather iron or generator types appears more logical. Thus, the fifth piece of predicted and confirmatory evidence (the ninth clue) favoring this revolutionary idea of a nuclear core in each of the bodies comprising our SS becomes more firmly established.

THE MYSTERIES OF EARTHQUAKES

The abstract (1992 Western Pacific Geophysics Meeting) *Infraplate Earthquakes and Crustal Horizontal Temperature Differences in Europe* by I. Stegena told of three series of tests that correlated the numbers of earthquakes in specific time frames with the geothermal temperature gradients in those areas.

In the Pannonian basin, temperature differences in 1 km depth are compared to earthquakes that occurred between 1859-1958. It was found that 95% of the earthquake energy in that time frame occurred on that half of the basin where horizontal geothermal gradients are large.

In the West European area, 84 of the 93 quakes between 1901-1955 occurred on that third of the area where the horizontal gradient of heat flow density are the most abrupt.

A third study carried out in East Europe gave a similar result in which 81% of the total quake energy between 1901-1973 burst out on only 20% of the area (where the horizontal changes of heat flow density are most abrupt).

The concordant results of these investigations, in which the epicenters are lying mostly on places of large horizontal temperature differences, suggest that the sporadic infraplate quakes are generated by thermal stresses and relaxation. The significant conclusion reached by Stegena is that there is no expressed correlation between tectonics and epicenters; the earthquakes of the area are not tectonic quakes sensu stricto.

At the same meeting in 1992, the abstract by Mary Ann Glennon on the subject of deep-focus earthquakes beneath the island of Sakhalin, reached the conclusion that at least part of the observed pattern residuals is due to path effect away from the source, not that of subducted slab.

The so-called 'rim of fire' virtually surrounding the Pacific Ocean conveys the notion to many minds that earthquakes and volcanic actions occur primarily along the boundary lines of two or more adjoining tectonic plates. However, there are too many exceptions to this assumption to make it a general rule. For example, during the week ending July 7, 1995, a *Chronicle Features* map pinpointed 14 significant earthquakes that had occurred during the previous seven days. Of this number, ten were inland, well away from known tectonic plate boundaries. Five of the ten quakes occurred within the USA.

In October 1992, Georgia Tech scientists announced that new analyses of molten, pressurized rock, minerals and other materials making up the mantle deep in Earth may provide insights into deep quakes, volcanoes and even the formation of the planet.

Scientists studying the mantle, which extends from about 60 miles to 1,800 miles below the surface, say the region consists of two distinct layers containing different proportions of key minerals. The difference between the two layers could help account for certain deep quakes and volcanoes that cannot be explained by conventional theories.

As far as they go, these findings and those of Stegena and Glennon are in complete agreement with the FLPM/IN concept. They, too, indicate the necessity of taking another look at current beliefs about the nature of earthquakes. The next step should be to question the source of the molten magna and its specific role in quakes. The answers will point the way to a better understanding of the nature and relationship of quakes and tectonic plates.

Most scientists now interpret their findings about quakes strictly in the perspective of plate tectonics in which rock slippage at the plate boundaries are the cause of the earthquake. In turn, these beliefs are linked to the Accretion Disk hypothesis, a supportive branch of the Big Bang theory.

However, when viewed in the perspective of the FLPM/IN concept, each booming epicenter of every quake *pinpoints* *a* *powerful* *explosion* that rocks the magma that causes the vicious or gentle shakings, cracking, sinking and uplifting of the crust. These events and the cracked eggshell-type surface they create worldwide attest to the powerful forces constantly at work deep within Planet Earth. This compelling new cause-and-effect version offers genuine opportunities for understanding the true nature of earthquakes.

Although not consistent in their warning signals, earthquakes do emit pre-quake clues: sudden changes in emissions of gases, electrical signals, and swelling of the ground under tremendous pressures. All of these signals can be traced back to the nucleosynthesis of matter in Earth's nuclear energy core.

THE WARNINGS OF PRECURSORY SIGNALS

The evidence for precursory warning signals continues to mount. An amateur who studies radio signals had warned that an earthquake capable of devastating damage would hit California within three days. Jack Coles, a former stereo salesman with no college degree, operates the Early Warning Earthquake Detection network out of his home in San Jose. The fax he issued on a Saturday in January 1994 said he had received reports of "increased radio signals, magnetic anomalies and many cases of electrical problems." He warned that the results could mean an earthquake measuring more than 6 on the Richter scale.

The quake that hit two days later in Southern California measured 6.6. Reporters had ignored Saturday's message.

Coles had started developing his theories a few years back when a radio he was repairing started making strange noises. "About four hours later we had a quake and I wondered if there was a connection," he said.

Scientists at the USGS in Menlo Park find little to back Coles' theories. "We sent three scientists over there some time ago to look at his stuff. We couldn't make any sense out of it," stated one scientist.

A study published in September 1992 suggested that the eruption pattern of an 'Old Faithful'-type of geyser (near Calistoga) in California could give warning of impending large quakes. In the report in the journal *Science*, scientists said they entered into a computer the record of the geyser eruptions since 1971 and then compared the eruption pattern for months around major quake events.

They found that three major quakes within 155 miles of the geyser occurred within one to three days after the geyser's eruption pattern underwent abrupt and dramatic changes. A mathematical examination of the Calistoga record eliminated both chance and the effects of rainfall as an explanation for the abrupt pattern changes preceding the quakes.

THE KOBE EARTHQUAKE SIGNALS

In January 1995, as aftershocks of the destructive Kobe earthquake continued to rumble beneath the region, reports of aurora-like flashes just before and after the deadly tremor were announced by a Japanese professor. Tamenari Tsukuda of Tokyo University said one of the more intriguing sights was a flash that streaked from east to west about eight feet above the ground shortly after the quake.

Phenomena like this are believed to be due to electrical and magnetic waves "generated by the grinding of Earth's crust," when interpreted in the "rock slippage" version of the cause of quakes. In the IN version, such phenomena are generated before, during and after the huge explosion at the quake's epicenter, while rock movements are the results of the powerful blasts.

Precursory signals that forewarn of pending quakes are very real, although not always consistent and not yet dependable. All signals can, and some day will, be traced to their source: Earth's nuclear energy core.

Radon concentration in ground water increased for several months before the Kobe earthquake. From late October 1994, the beginning of the observation, to the end of December 1994, radon concentration increased about fourfold. On 8 January, nine days before the quake, the radon concentration reached a peak of more than ten times that at the beginning of the observation, before starting to decrease. These radon changes apparently are precursory phenomena of the disastrous quake.

Chloride and sulfate ion concentrations of ground water issuing from two wells located near the epicenter of the Kobe quake fluctuated before the magnitude 7.2 event on 17 January 1995. The samples measured were pumped groundwater packed in bottles and distributed in the domestic market as drinking water from 1993 to April 1995. Analytical results demonstrated that the concentrations of both ions increased steadily from August 1994 to just before the quake. Water sampled after the quake had much higher ion concentrations. The precursory changes in chemical composition apparently reflect the preparation stage of a large earthquake.

These precursory changes in concentrations of radon and other chemicals do not appear to have any connection with strain buildup in rocks. Rather, they appear to lead in a direct path to internal nucleosynthesis as the source of the changes and the basic cause of earthquakes.

DEPTH AS A SAFETY FACTOR

On June 8, 1994, a powerful magnitude earthquake sent panicked Bolivians, Chileans and Brazilians into the streets and was felt as far away as Canada. Power outages occurred in parts of Chile and Bolivia. Although it was perhaps the biggest deep-focus quake of the century, there were no reports of major damages or casualties.

Unusually deep at 400 miles beneath Earth's surface where solid rock does not exist, the widely felt quake caused very little damage. The lack of rocks there allowed the shock waves to travel through the magma and far into the distance, dissipating the explosive energy quickly and safely.

Scientists have learned that our Sun has radial pulsations that make the solar surface contract and expand like a ringing bell. And like a bell, Earth has its own natural frequencies -- or normal modes -- which start ringing if the globe is hit hard enough -- as was the case here. The most persistent of these modes causes the planet to expand and contract every 20 minutes, almost as if it were breathing. This mode can be detected even three months after a great quake. "This is really going to change the level of our information about the deep Earth," one seismologist said.

Contrary to this magnitude 8.2 quake, the much smaller, shallower magnitude 6.7 quake in Northridge, California killed 61 people and caused at least $20 billion in damage.

Scientists have yet to unravel exactly the cause of deep earthquakes. According to one prominent, but speculative theory, deep tremors occur when increasing pressures cause minerals in the ocean crust to undergo a sudden structural transformation!

Such speculation is not necessary; the cause is obvious. In any size quake, the deeper the explosion's epicenter, the less the danger at Earth's surface. This holds especially true when deep explosions occur where no rocks exist. If no rocks exist there, then rock slippage cannot be the cause of the quake. If not the cause, crustal rock movement must be the result of a powerful explosion at the epicenter of each quake, no matter how deep or shallow. If below the level of solid rock, Earth's bell will ring until the shaken magma settles down again.

WHY THE EXPLOSIONS?

In April 1992 an explosion packing the power of an earthquake ripped open an underground propane gas pipeline, killing one person, flattening nearby mobile homes and shaking buildings more than 140 miles away. It registered 3.5 to 4.0 on the Richter scale -- as strong as an earthquake that could cause slight to moderate damage.

"It was just a big bang, a tremendous bang," stated one survivor. One child was killed, and at least 18 persons were injured, three critically.

The explosion occurred in a rural area seven miles south of Brenham, Texas, a community of 12,000 about 70 miles northwest of Houston. Officials suspect gas at a liquid petroleum storage and pumping facility collected in a ravine and was ignited by a car or a pilot light in a home.

A most interesting point here is the comparison of this powerful explosion to a 3.5 to 4.0 quake. One must wonder how different or how much alike the two events actually are.

THE TINY MYSTERY OF POLONIUM HALOS:
Creationism, Big Bang or the FLPM/IN?

In the book *Creation's Tiny Mystery*, Robert V. Gentry presents a good argument for Creationism, using polonium halos found in granite as evidence supporting this belief in opposition to the evolutionary viewpoint of science. The book was sent to me in June 1995 by Glenn C. Strait, Natural Science Editor of *The World&I* magazine. His cover letter asked how the presence of these tiny halos in granite could be explained by the FLPM/IN concept.

Under the microscope these halos show a tiny radioactive particle at the center of concentric ring patterns in the granite, much like the bull's eye at the center of the rings. Because of their radioactive origin and their halo-like appearance, these microscopic ring patterns became known as radioactive halos.

After reading the interesting book, the essence of my reply can be summed up in two short paragraphs: While I do not agree with the conclusion that his discoveries of polonium halos in granite support Creationism, I am excited about two aspects of these findings. First, *they offer strong support for the evolution of planets via internal nucleosynthesis (IN), which, in turn, is powerful evidence against the Accretion Disk hypothesis of the origin of the SS* (a vital factor in the Big Bang theory).

Secondly, the short half-life of radioactive polonium gives solid assurance that they could not, by any stretch of the imagination, have been formed in distant supernova explosions (as claimed in the Big Bang theory) and survived the eons that supposedly elapsed before they became a part of Earth's crust. Thus, in the prevailing beliefs about planetary origins, it is impossible for polonium to be a primodial constituent of Earth's granite.

The evidence clearly shows that the halos had to have been formed in situ -- and they were. *Only in the FLPM/IN concept would this be possible!*

Both the origins of Precambrian granite and the polonium halos therein *can be explained readily in this IN perspective* -- a concept firmly structured on the solid foundation of the SS's geometric origin that proved crucial in deriving the Fourth Law of Planetary Motion on the spacing of planets.

To expound further, granite is the foundation rock of Earth's ever-thickening crust. It was among the first layers to solidify atop the molten mantle worldwide to form tenuous sections, and eventually a permanent foundation on which other layers gradually built -- and are still building. How did the compositional elements and compounds form into granite with polonium halos inside?

Polonium is a radioactive metallic element belonging to the uranium decay series. It occurs naturally in pitchblende as a decay product of radium, and can be produced artificially by bombarding bismuth with neutrons. Its most stable isotope has a mass number of 210. Polonium, with a half-life of about 138 days, decays into an isotope of lead by giving off alpha rays. One polonium isotope is the product of the radioactive decay of radon, a common gas that still emanates continually from deep within Earth.

The short half-life of polonium-218 of three minutes means that every three minutes, one-half of its remaining mass will decay. If created, along with other elements, within Planet Earth, it would be no surprise that traces of any and all polonium isotopes are found in granite. The isotopes formed in situ, during cool down of the granite mix containing the polonium and other elements.

The textures and composition of granite give important clues to its formation processes. Scientists know that these foundation rocks have coarse-grained, crystalline textures which are only found in rocks that cool slowly from a hot molten mass. The process can be observed by crystallizing compounds in the laboratory. These experiments always form crystalline textures.

These clues point to the only apparent explanation: the elements and compounds comprising Earth's hot mantle-like, thick liquid surface some 4.6 billion years ago came together and slowly cooled to form granite. Radioactivity was at its peak surface performance; huge numbers of large particles of radon, polonium, etc. continually bombarded the coagulating granite; particle entrapment was common, followed by decay that left their halo marks in the granite.

Through the eons the newly-mixed materials from the hot mantle continuously contributed to the ever-thickening crust. The process continues today, and will do so at diminishing rates until all the energy of Earth's core is expended in the process called nucleosynthesis.

Thus, to the previous nine clues to Earth's nuclear core, we can add one more: the tiny mystery of polonium halos in granite. While adding much support for the FLPM/IN, these tiny clues to Earth's origin clearly aim their arrows at both the hearts of Creationism and the Big Bang theory.

Just before this book went to press, a stunning weather report from the probe launched into Jupiter's atmosphere (7 Dec 1995) by the Galileo satellite hit the news media. The report revealed that the skies were hotter, windier, drier, and clearer than forecasters predicted. In fact, the prelimary data from the probe is so shockingly different than expected that it inevitably will lead planetary scientists to rethink not just the meteorology of the gaseous planet, but its very origins. According to the understatement of one Galileo project scientist, the data "doesn't fit very well. In fact, it's darn uncomfortable."

Contrary to this situation, the report's results are very exciting in that they are precisely what one would expect when interpreting the data in the perspective of the FLPM/IN concept. The best way to verify this statement is by comparing the intepretations of the data in the two different perspectives: the current view of Jupiter's origin versus the FLPM/IN view.

The temperature at Jupiter ranged from -144°C at the top of the ammonia-cloud covered atmosphere to +152°C (306°F) at only 600 km into the thick blanket of clouds covering the giant planet whose diameter measures 142,980 km. Pressures ranged from 400 millibars to 22 bars over the same descent path, compressing the gases to densities up to 100 times greater than previously postulated. The extremes of temperature and pressure created a vertical convective motion in the atmosphere, stirring up turbulent winds more than 50 percent stronger than predicted. These winds were fairly constant throughout the probe's descent.

In theory, the probe should have passed through a region where wind speeds drop to zero. But the it never reached such a point. Contrary-wise, the wind speed increased with depth, which led several investigators to speculate that the energy source driving the circulation of the Jovian atmosphere is probably escaping from the interior. That answer is a giant step closer to the truth.

In the perspective of the FLPM/IN concept, one would predict conclusively that the engine driving Jupiter's powerful winds is the internal heat source: its nuclear energy core. The core also accounts for the thick, extremely dense blanket of gaseous clouds and the high temperature at the depth of only 600 km. Jupiter's core temperature, even though well insulated, was recorded by Pioneer 10 in the 1970s at 30,000°C, but this will prove to be a very conservative figure when more accurate measurements can be made.

The probe's helium abundance detector recorded that the outermost regions of Jupiter now contain much less helium than the planet started with, a figure calculated from the helium-to-hydrogen ratio in the Sun. Current beliefs go with the suggestion "that Jupiter's helium is now condensing into droplets under the deep interior's megabar pressures; the droplets then fall even deeper into the planet. So the gravitational energy released as heat by the fall of the helium raindrops must in fact be fueling Jupiter's infrared glow, which is brighter than anything the solar energy reaching the planet could account for."

A more sensible explanation is that Jupiter's core, in strict compliance with natural laws of evolution from energy to matter, simply is producing a smaller percentage of helium as it gradually evolves into the stage that produces less helium and/or more heavier elements. Rather than being produced by helium raindrops, the heat fueling Jupiter's infrared glow emanates directly from the nuclear core, just as predicted by the FLPM/IN concept.

Further, the probe revealed Jupiter to be much drier than anticipated and relatively free of condensation. The observation that its atmosphere contains water concentrations equal to that of the Sun left investigators wondering. However, both the helium and water concentrations are in line with predictions of the FLPM/IN concept in that they reveal the close relationship of the two masses. They are clear indications that Jupiter has evolved just recently from the energy stage by covering itself with the initial blanket of chemicals that still match the Sun's production. As the huge planet evolves toward the rocky stage, its production of chemicals will gradually lean more and more toward the heavier elements, and its higher-than-expected radiation doses will gradually decline as more and more of its core energy is transformed into matter.

Scientists can learn a great deal about planetary evolution through the realization that planets actually do evolve from energy masses through five common stages. Jupiter is an excellent specimen to study for details of evolutionary changes as the giant planet progresses through the second stage.

As for water clouds, mere wisps of water particles were found in the atmosphere. Even many of the heavier elements -- carbon, oxygen, and sulfur -- as well as neon were found to be at lower-than-expected concentrations. Here again, these findings are revealing great details about planetary evolution. There simply is no need to use the excuse that Galileo's probe dropped into a very rare "hot spot" to explain these unexpectedly stunning results. With winds at 531 km/hr, temperatures would tend to level out any such spots. It a safe bet that the results are truly representative of the atmosphere on Jupiter.

The two remaining years of observations to be made by the Galileo orbiter should result in many more exciting discoveries that will be confirmatory evidence of the FLPM/IN concept of planetary origins.

Just as Galileo himself drove the dagger into the heart of the Ptolemaic version by promoting the Copernican idea of a Sun-centered Solar System, the Galileo orbiter is plunging a dagger into the heart of the prevailing dogma about planetary origins.

ABSTRACT: All efforts since the initial attempts of Johannes Kepler in 1595 to solve the mystery of how the planets attained their orbital positions around our sun have failed to produce a definitive solution to this enigma. To understand the mystery of the origins of solar systems and the evolution of planets, one must first explain, beyond doubt, the critical relationship between the current spacing of our planets and the forces that power the evolution of these celestial spheres. Prevailing concepts fail to explain this inseparable connection.

The key to understanding this relationship resides in Kepler's First Three Laws of Planetary Motion and the solution to the proposed Fourth Law of Planetary Motion. Together, these Four Laws offer a valid explanation of the dynamic manner in which our Solar System came to be, while providing a solid foundation for understanding the forces that drive the evolution of planets. The complete mathematical solution is detailed.

A valid solution must provide a solid foundation for supplying definitive answers to all anomalies of our solar system; e.g., how and why planets attained their current orbital positions, why planets differ in size and composition, how and why the atmospheric and surface features of every planet progressively undergo evolutionary changes, and what forces drive planetary evolution. Why does Earth contain the full range of elements from hydrogen to uranium, while only the lighter elements can be found on Jupiter, Saturn, our Sun, extra-solar planets, etc.? How and why did the   known extra-solar gaseous giant planets form so close to their central star? Why will water and signs of early forms of life likely be found on each of our nine planets and many moons?

The corroborating evidence is both substantive and substantial. A definitive and testable alternative, the revolutionary FLINE model of planetary origins and evolution is brought full circle by this solution to the proposed Fourth Law of Planetary Motion, now undergoing peer review. The new model reveals the crucial relationship between the spacing of planets and the forces that power their ongoing stages of evolution.

Principles of the FLINE Model

The solution to the 400-year mystery proved to be the final link in a revolutionary model of the origin of our Solar System and the evolution of planets -- a new concept that evolved during the research years of 1973-1995. According to Scarborough, "This geometric solution that first eluded Kepler in 1595 proved to be the key to bringing the decisive discoveries of the past millennium full circle. It all began with the realization in 1973 that hydrocarbon fuels (gas, petroleum, coal) could not have been created from fossils. The encapsulated imprints of live plants and animals found within coal -- the last stage of polymerization of these three fuels -- could have been preserved only via a sudden encapsulation of the live plants and animals by petroleum gushing from the earth and inundating lowlands, then polymerizing and cross-linking as solidified coal via these very common processes of Nature. Substantive evidence since then has confirmed beyond doubt that these fuels were created -- as was all matter comprising Earth -- via these natural processes of Nature, and not from fossils."

The ongoing creation of atomic matter comprising all planets would not be possible without an internal source of energy that keeps the planet active until the energy source is depleted. These internal processes eventually push each planet through five common stages of evolution. Earth is in the fourth (rocky) stage; Mercury and Pluto are in the fifth and final (inactive) stage.

Scientists know that all stars are driven by these internal processes called internal nucleosynthesis (i.e., energy into matter, in accord with Einstein's formula $E = mc^2$). Many scientists also realize that all planets of the Universe are driven via internal nucleosynthesis, often referred to as the heat within. However, even though the Four Laws of Planetary Motion (FL) reveal the fiery, dynamic origin of our Solar System and that the internal heat, properly called internal nucleosynthesis (IN), is the driving force behind planetary evolution (E). These three principles of the FLINE model are interconnected, inseparable, ongoing essentials that drive all planets through their 5-stage cycles of evolution. We have the energy-matter transformations known now as internal nucleosynthesis, he did realize that all planets evolve from energy masses into planetary spheres. His writings on the subject appeared 30 years after Johannes Kepler had discovered the last of his three laws of planetary motion, but had failed in efforts to discover the fourth law explaining the geometric spacing of the sixth then-known planets. Soon afterwards, Newton discovered gravity holding the solar system together. These great discoveries, along with those of a number of geologists, form the solid foundation of a revolutionary definitive concept of planetary evolution.

The discovery of the mathematical solution to the Fourth Law during the years of 1980-1995 proved to be the final link in the new model of the origin of our solar system and the evolution of planets through five stages of evolution via internal nucleosynthesis. The Four Laws (FL) clearly reveal the dynamic origin of the solar system and how the planets attained their orbital spacing. The processes of internal nucleosynthesis (IN) in full accord with Einstein's formula clearly are the forces that drive all planetary evolution (E) forward until the energy-core source is depleted. These three interlocking and inseparable principles

serve as a solid foundation for the revolutionary FLINE model of the planetary origins. This new definitive concept provides crystal clear explanations that are continually corroborated by every relevant discovery about Earth and all other planets, including the recently discovered exoplanets. The solution to every anomaly of every planet and moon has its taproot deeply embedded in the Fline model--a statement already proven true by its record of accurate predictions.

The new model remains poised to displace the prevailing, but highly speculative, Accretion concept that has survived via speculative modifications since its introduction in 1796 by Pierre Simon Laplace in his Exposition of the System of the Universe. Einstein's work defining the energy-matter relationship that earned him the title of Man of the Century, along with Descartes' insight into the nature of planetary cores, Kepler's Three Laws of Planetary Motion and the new Fourth Law stand as beacons capable of guiding scientists to a fully proven understanding of the origins of solar systems and the evolution of planets--and perhaps to understanding our Universe--in the 21st century.

Fusion-fission processes occur in all spheres of the universe; in making the stars shine, they are the universal powers that drive forward the evolution of all spheres. Rene' Descartes, brilliant in his own right, first recorded in 1644 that all planets began as spherical masses of "Sun-like" matter, and while unaware of

Sadly, the disbelief in Darwin's animate evolution concept has an equally tragic parallel in the inanimate evolution of all things universal. The two are intimately intertwined; both stem from the same taproot of origins: energy. Nature's greatest universal truth can be summarized in two words: *everything evolves*. Evolution -- animate or inanimate -- is not possible without an energy source to drive it forward. Every planet is a self-sustaining entity that creates its own compositional features via the internal nucleosynthesis of atomic matter and subsequent polymerization, etc -- including any form of life existing thereon. We do not need to look to outer space for answers; they have always been right before our eyes.

The definitive FLINE model of planetary origins (1973-1999) consists of three chronological, inseparable and ongoing realities of Nature: (i) the Four Laws of Planetary Motion (FL), (ii) internal nucleosynthesis (IN) and (iii) evolution (E). No solar system is possible without undergoing these inseparable realities that remain interactive until one by one, they reach their ending. They reveal that planetary evolution is not possible with any type of core other than that hypothesized by Descartes as "Sun-like" -- a concept now corroborated by abundant substantiated evidence. Such evidence mounts with each relevant discovery of planetary anomalies, both of Earth and of space probes. From energy mass to inactive sphere, every planet is a self-sustaining entity that creates atomic compositional matter via internal nucleosynthesis throughout its first four stages of evolution. We have only to look to the skies to observe each and every one of the five stages of planetary evolution and their relationship to mass.

Whether speaking of the animate or the inanimate, the evidence for evolution is indisputable, incontrovertible, overwhelming and conclusive. To the two revolutions Freud designated as paramount (the ideas of Copernicus and Darwin), we must add another one that is destined to displace prevailing beliefs: the FLINE model of planetary origins and evolution. Had Kepler succeeded in his quest (initiated in 1595) for the Fourth Law of Planetary Motion, there would have been no need for scientists later to establish the speculative, but erroneous, concept of planetary accretion from gases, dust, asteroids, comets, etc. The solution* (1980-1995) to the elusive Fourth Law was the final link in the FLINE model that inevitably, as with the ideas of Copernicus and Darwin, will force scientists to rethink their beliefs about inanimate and animate origins. In view of its factual, non-speculative structure and vast scientific potential, why must this new model face the same hurdles encountered by those revolutionary ideas?

CONCLUSION

How can the following be explained except via the FLINE model of planetary origins?
1. The several iridium layers found on Earth.
2. The titanium layer found on the Moon.
3. The tellurium layer found on Venus.
4. The magnesium layer found on Neptune.
    [Did each (or any) of the above four types of layers selectively come from distant supernovas?
    Or did a different type of comet selectively crash on each one of them? If so, why did the sever-
    al iridium comets always crash only on Earth?]
5. The abundant (virgin) metals observed at the impact site of fragment G of Comet SL9 on Jupiter.
6. The atmospheres and surface features of planets and moons, and why they differ in accord with size.

The answers are easy to understand when we realize that all are in situ ejecta products of IN; these things did not, and do not, come from outer space. If only scientists were given the opportunity to learn about the FLINE model, there would be no need to explain everything by simply saying that it came from outer space long ago and far away.

Factually based and without the need for speculation, the revolutionary FLINE paradigm of planetary origins and evolution is structured on three inseparable and indisputable principles of Nature: the Four Laws of Planetary Motion (FL) and the internal nucleosynthesis (IN) that drives evolution (E) in all active spheres of the Universe. Imbued with these qualifications, the FLINE model appears destined to become one of history's most significant and crucial scientific breakthroughs in the planetary sciences.

By firmly interlocking the great discoveries of Copernicus, Galileo, Kepler, Newton, Descartes, Dutton, Einstein and many others, this new model stands alone in bringing their significance full circle. Other known models neither establish these critical connections nor are capable of explaining every planetary anomaly, including those of the exoplanets. Scientists will find that every discovery -- past, present, and future -- has its taproot deeply embedded in the principles of the FLINE model. It is the only model in which its every facet can and will be proven beyond doubt, and the only model in which no fundamental flaw has been or can be found. In its capability for explaining every anomaly of solar systems, the new model stands alone and shines a powerful spotlight on the absurdities of the Accretion concept of planetary origins -- an exposé of immense implications.   The great potential for huge savings in research time and money could be realized if scientists would be willing to examine their findings in the perspective of this revolutionary FLINE model of planetary origins.

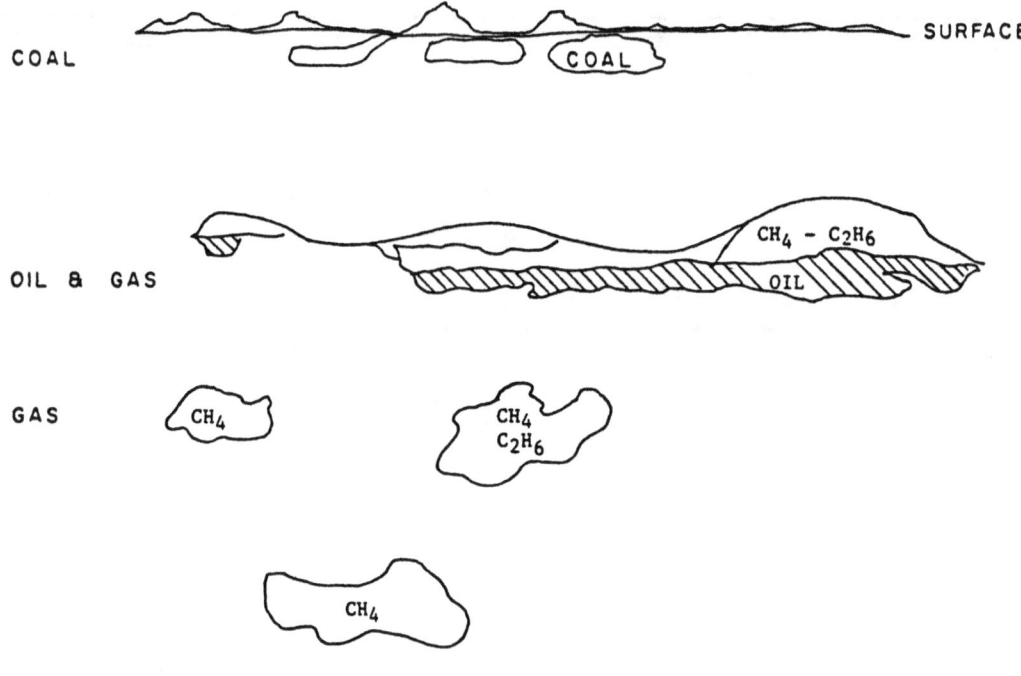

COAL

SURFACE

COAL

OIL & GAS

$CH_4 - C_2H_6$

OIL

GAS

CH_4

CH_4
C_2H_6

CH_4

LAYERED HYDROCARBON FUELS

Coal, petroleum and gas are generally layered in the descending order shown.

Next, the manner in which Earth's systems evolved will be explored. The drawing above is intended to fix in the reader's mind the three-layer system of hydrocarbon fuels in Earth's crust. Understanding how the gas, petroleum and coal formed leads to easier comprehesion of the manner in which all systems of all planets formed and are still forming.

68

CHAPTER III

HOW EARTH'S SYSTEMS EVOLVED

THE MYTH OF FOSSIL FUELS

The third and final phase of this FLPM/IN concept describes how planetary systems evolve from energy into the matter that makes Earth such an ideal planet for all of us.

The example of the origin of hydrocarbon fuels (gas, petroleum, coal) best illustrates the processes whereby Nature makes all things by means of natural laws. A little background is essential.

During the 1830s, William E. Logan, a graduate of Edinburgh University, managed his family's coal and smelting interests in Wales. Logan's great interest in the origin of coal led him to study some 100 coal beds and seams. In every case he made note of three observations:

1. A bed of bleached clay lay under each coal bed.

2. Within the clay beds were tangled masses of long, slender, fibrous root systems with a thin coating of carbonaceous matter.

3. Well preserved imprints of ferns and other plants were scattered throughout the coal.

Logan concluded that plants, specifically the Stigmaria ficodes, had turned into coal. Stigmaria structures, some microscopic, some full size, plus larger fossils (branches and tree trunks) were found in the coal. He recorded that "in Stigmaria ficodes we have the plant to which the Earth is mainly indebted for those vast stores of fossil fuels."

Then in the 1920s, J.B.S. Haldane's hypothesis of petroleum created from tiny marine organisms added more credibility to the concept of fuels made from fossils. So it seemed logical that all natural gas is a product of decomposition of plants and animals. Thus, the fossil fuels theory (FFT) became firmly entrenched in scientific literature.

Since Logan's time, advocates have attempted, in difficult and costly efforts, to explain their relevant discoveries in its perspective. In reality, as we shall see, much time, frustration and money could be saved if scientists have opportunities to interpret their findings in the new perspective of the FLPM/IN concept.

Contrary-wise, through the years, there have been a number of opponents who kept faith in the belief of non-biological (abiogenic) origins of fuels. The list includes some historical names: Berthelot (1866), Mendele'ev (1877, 1902), Humbolt, Gay-Lussac and others.

In more recent times, a significant volume of evidence that argues against the validity of the FFT has accumulated in scientific literature. Through intensive library searches, the author has gleaned and condensed much evidence into six critical points:

1. The tremendous volumes and extreme depths of fuels (especially gas) contradict, rather than enhance, the FFT.

2. The patterns of distribution, size and thickness of coal beds and seams simply do not fit into the FFT.

3. No multi-layers of root systems are found in coal beds; carbonized root systems are found only in under-bed clay.

4. Plants yield vegetable oils; coal and petroleum are mineral oil based.

5. Peat, credited as a transitional stage of some plant-to-coal processes, actually turns into black dirt that retains no imprints of Stigmaria ficodes.

6. The structural integrity of plant imprints can be preserved only through encapsulation of live plants before decay begins.

Each point adds weight to the powerful evidence accumulating against the FFT. For example, today we know that a fossil imprint found in rock was still in its original life form when encapsulated by softer material that later solidified around it. The imprint remained in the rock after dissipation of the encased specimen, which obviously cannot be credited with creation of its encapsulating rock. In this perspective, we can reasonably conclude that the plants whose live imprints were discovered by Logan could not have created the coal; they simply were victims of encapsulation.

Logan erroneously concluded that coal was made from plants. His three astute observations were misinterpreted. In reality, they argue strongly against his hypothesis. Such integrity of plant structure could not have been preserved via the decaying leaves interpretation made by Logan *et al.* Further, his findings can be explained more reasonably by the FLPM/IN concept. Only in this perspective can his three observations be interpreted accurately.

Only hot, chemically-contaminated petroleum could have bleached the clay beds and seeped down the root systems to preserve them via carbonization. Only the sudden encasement of live swamp plants by an encapsulating medium (petroleum) could have preserved their live structural integrity, discovered as imprints of live plants. How did it all happen?

HOW HYDROCARBON FUELS FORMED IN EARTH'S CRUST

As confirmed in a research project in Neely Nuclear Research Center at Georgia Tech in 1985-6, the three hydrocarbon fuels (gas, petroleum and coal) that abound throughout Earth's crust, generally exist in overlapping three-layer systems. Coal is the surface fuel, found on the surface or within the top mile of the crust. Petroleum dominates farther down at medium levels, while methane gas is found vastly more abundantly at deeper depths.

At the center of Earth lies its nuclear core, the driving source of the energy that is transformed into the atomic elements that comprise the hot material of the mantle from which the crust is being formed continually. Carbon and hydrogen are among the most abundant elements created in the nucleosynthesis processes. They are the building blocks of methane gas; the two elements combine to form the vast quantities of methane gas found throughout Earth's crust.

During the rocky stage of evolution, virgin elements combine in similar manner to form a large variety of compounds that become the crust and the planetary systems comprising Earth's ever-thickening shell. In our fuels example, much of the methane links together, through the process known as polymerization, to create other gases: ethane, propane and butane.

When five of the methane molecules linked together, they formed the next product in the carbon-chain series: pentane, the first liquid and the lightest component of light petroleum. Over time, polymerization continued to forge larger and larger molecules of thin and medium weight oils, and finally, thicker crude petroleum. While unknown to Logan *et al.*, scientists now know that all crude petroleum contains tiny particles of coal -- the beginning of Nature's final transitional phase from gas to petroleum to coal.

When huge quantities of crude petroleum were forced onto Earth's surface, large areas of swamplands were inundated with the hot encapsulating oils that eventually polymerized, cross-linked and solidified as coal.

From this evidence, we can reasonably conclude that the hydrocarbon fuels, gas, petroleum and coal, should be called 'energy' fuels rather than 'fossil' fuels.

Strong evidence supporting this conclusion abounds. From the scientific literature, the five recognized facts that dramatically reveal the very close relationship of the three energy fuels can be summarized as follows:

1. All wet gases contain lightweight oils (condensate; e.g., pentane), illustrating the first transitional phase via polymerization from gas to petroleum.

2. All petroleum contains gases in varying amounts inversely proportional to the degree of polymerization.

3. All crude petroleum contains tiny particles of coal, illustrating the second transitional phase via polymerization from petroleum to coal.

4. Gummy coals (boghead and cannel) can be classified as either petroleum or coal, illustrating an advanced stage of the second transitional phase (additional polymerization and some cross-linking).

5. Every lump of coal contains oils and gases, illustrating all three phases and the degrees of polymerization and cross-linking of the three fuels, while confirming their close relationship.

A close study of the evidence clearly reveals that gas, petroleum and coal are the first, second and third phases in the formation of hydrocarbon (energy) fuels, and that polymerization is the key to evolutionary changes from gas to petroleum to coal.

THE THREE-LAYER SYSTEM IN EARTH'S CRUST

Why would coal, if formed from great swamps of plants and trees, generally be found among and under the rocky layers of mountainous regions like West Virginia? Why are coal beds located at or very near the surface? Why is coal found in veins, many of which are small in diameter and relatively short in length? Would not the predominantly mountainous distribution pattern of coal and the shapes and sizes of coal veins indicate that great forces pressured gases and liquids into such locations and formations, where polymerization to petroleum and solidification to coal occurred? If so, how could one explain the processes, in terms of natural laws, that created these situations?

Why does petroleum exist generally at intermediate depths ranging from near the surface down to 30,000 or more feet? Why is it that the deeper the drilling of wells, the greater the proportion of gas to petroleum? Why should gas be the deepest of the three hydrocarbon fuels? Why do no fossils exist in the deeper oil and gas sites? And how did gas form at such great depths in unimaginable quantities under tremendous pressures at such excessively high temperatures?

Through literature searches in many libraries over the ensuing years, confirmations of the imagined answers to the persistent questions rattling in my brain were sought and found. The excitement of discovering in the literature the facts that substantiate the wild imagination serve to spur one on to the next phase of the problem. There can be no turning back or termination of the quest for truth: one can become obsessed with it. This, in spite of the discountenance certain to be encountered along any way that is in opposition to established beliefs. While that can be traumatic, it cannot be a deterrent to progress. History is replete with such examples, and time will never alter the cycle.

To summarize briefly, the energy fuels are arranged generally in three overlapping layers in Earth's crust. Coal is found on or near the surface, while petroleum is found at lower (medium) depths. Deeper drilling results in higher ratios of gas to petroleum. Finally, the deepest drillings may yield only gas. These overlapping layers of gas-to-liquid-to-solid fuels are arranged roughly in ascending order from Earth's interior to its surface by natural laws of physics and chemistry.

The first publication of the three-layered system of fuels in 1975 (*Fuels: A New Theory*) by the author understandably met apathy and some opposition from the disciples of the FFT. Additional publications in 1977, 1979 and 1986 managed to gain a few converts, but the numbers remain small at this writing.

Meanwhile, in February, 1981, an article on *World Energy Resources* was published in a special report in *National Geographic*. Included were three maps of the North American continent showing the distribution pattern for each of the three fuels. When superimposed, the three maps revealed very similar distribution patterns for gas, petroleum and coal over the whole continent, thereby adding much credibility to the three-layer concept.

With this background, let's take another look at Logan's three astute observations of the 1830s, and, working in reverse of Nature's processes, continue developing the new theory of formation of these hydrocarbon fuels.

HOW COAL FORMED FROM PETROLEUM

When Logan saw the bleached clay under every bed of coal, he should have questioned why it was bleached. Bleaching is usually accomplished by a combination of certain chemicals at high temperatures. Imagine what might occur if large volumes of petroleum were forced from the ground under thousands of pounds pressure and at very high temperatures. Such a familiar gusher happened in 1979 in the Gulf of Mexico when an oil well flowed threateningly for nine months before it was forcibly closed off.

Hundreds of millions of years ago, such hot oil erupting onto land filled all the low areas (generally swamplands) within reach. It inundated low-lying areas where Stigmaria was usually the dominant plant life. The high heat and chemicals in the petroleum bleached the clay underneath, thereby leaving the first clue for Logan to discover.

By seeping downward and penetrating the root systems of plants, the oil preserved the "tangled mass of long, slender fibrous root systems with a thin coating of carbonaceous matter" in their original form. The carbonaceous matter found by Logan were the remains of the oils that had preserved the root systems. If not for this preservative, the root systems would have disappeared rapidly from the scene, as all roots normally do.

Ferns, leaves, branches and even tree trunks were scattered, suspended throughout the lake of hot viscous oil. Thus they were preserved in situ, destined to leave their distinctive imprints, porphyrins and carbonized skeletons in the mass as the oil cooled, thickened, solidified, cross-linked and polymerized into solid coal. And thus time, temperature and pressure inevitably changed the oil channels and deep ponds into coal veins and coal beds, some as much as hundreds of feet thick. Much of the gas and oil was forced by the tremendous pressures into buried crevices, deep sand formations, porous rocks and strata where it transformed into vast quantities of shale oils and tar sands.

Thus the world's coals were created when extremely hot petroleum containing gases from Earth's interior poured out over the swamplands, encasing the plants, seeping down the root systems, and bleaching the under-beds before solidifying into solid coal.

But what could be the source of such vast quantities of petroleum? Could these oils have formed from deeper gases that polymerized?

HOW PETROLEUM FORMED FROM GAS

If coal formed from petroleum, then the next step is to find the source of the oils. Nature creates petroleum by the process of molecular chain-building; i.e., methane gas molecules are joined together, much like links in a chain, by natural processes (polymerization) into longer, larger molecules known as light oils, or condensate. The lightest of these is identified as pentane, because each molecule contains five atoms of carbon. Each time another molecule of methane is polymerized into the chain, the larger molecule is renamed in ascending order: hexane, heptane, octane, etc. Various proportions of these condensates always are found in all wet gases and in all lightweight petroleum.

As polymerization continues, the oils become heavier and thicker, more viscous. At a higher viscosity, the petroleum becomes known as a crude oil. During these transformations, a process called cross-linking is initiated to make ever larger and more rigid molecules. While the term is self-explanatory to many readers, its meaning should be clarified: the molecular chains link together side-to-side, thus becoming less mobile, more solid. Finally, when sufficient polymerization and cross-linking occur, the petroleum begins to solidify into the tiny particles of coal always found in heavy crudes.

Where and why does polymerization of gas into petroleum occur? Starting deep inside Earth, methane gas, under tremendous pressure, seeks the paths of least resistance on its upward journey. Most of it becomes trapped either in porous rock or strata or beneath impervious barriers. Here the high heat and pressure, in conjunction with the contaminants (trace metals) in the gas, initiate and sustain the well-known chemical process of polymerization.

During the past two decades, duplication of the process has been accomplished in a number of chemical labs, and patents have been issued covering the catalytic conversion of methane into gasoline (a mixture of lighter weight oils). The process is not commercially feasible at this time.

A fantastic example of such lightweight oils created by natural processes was discovered on June 5, 1989, near a town south of Riyadh called Hawtah in Saudi Arabia. The gusher tested at 8,000 barrels a day. Later, four more prolific oil wells and one gas well were drilled there. The vast deposits contain a rare, superlight oil that could be used in a car engine without being refined. The new oil has the color of gasoline and the consistency of water. Its gravity varies from 42 to 49 degrees, as measured by the standard API method. The Hawtah oil contains almost no sulfur or other impurities. The oil field may prove to be the world's largest.

Only a week before Iraq invaded Kuwait, Aramco announced the size of the prospective drilling area: 1,440 square miles. The discovery raises questions that may never be answered: Was this field the next target envisioned by Saddam Hussein? Was it the primary reason he gave up the objectives of the 8-year war to Iran and turned his full attention southward? Was his primary objective the control of oil prices worldwide?

One would expect the Hawtah oils to be free of the contaminants found in crude oils, and they are. Petroleum crudes usually contain numerous substances, both organic and inorganic, including many trace metals, salts, lignite and coal. Microscopic studies reveal fragments of petrified wood, spores, algae, insect scales, tiny shells and fragments. In contrast to Haldane's theory, it is safe to conclude that these substances are contaminants trapped in the oil; they did not create the oil! These contaminants remain in the mass during and after its transition into coal, thereby often participating in the reactions that create the many chemicals found in coal.

It is important to remember that not all gas polymerizes into petroleum and coal. Molecules of gas are found in every lump of coal and every drop of petroleum. Pockets of these gases in coal mines, when released from entrapment in the black gold, present great dangers of explosion and suffocation.

SOURCE OF THE METHANE GAS

If coal was made from gushing petroleum that had been created from deeper gas, what is the source of the methane? How was it made in such unimaginable quantities under such extreme conditions of temperature and pressure? Is it still being created? The answers to these questions are profound, revealing a logical concept in which the creation of all matter comprising our planet can be explained with very little, if any, speculation.

Before delving into these questions, we need to take a closer look at the vast resources of gas in Earth's crust. Geopressured methane, an example of one type of gas, lies in the untapped gas-laden briny waters buried deep beneath the Gulf Coast. A Baton Rouge hydrogeologist, a leading authority on the subject, calculated the supply at 50,000 trillion cubic feet (tcf) of methane in Louisiana and Texas. The annual use rate of 20 tcf in the USA equates to 2,500 years of reserves in this one location. One tcf is the energy equivalent of approximately 180 million barrels of oil.

In addition to the Gulf Coast reserves, methane is known to exist in a large basin area some 500 miles in length in California, and in other States, including Oklahoma, Washington, Oregon and Alaska (250 tcf). These resources can be multiplied many times in the worldwide perspective, since there is no reason to believe that the listed areas have a monopoly on gas supplies. As in past decades, even these estimates will prove to be far too conservative.

So we are talking about unimaginable quantities of methane inside Earth -- stupendous amounts far too deep and too voluminous to have been made from fossils, or to have come from outer space. What might be a logical source from which such huge volumes could evolve?

One clue lies in finding the source of the carbon and hydrogen atoms that combine to make methane. How and where can these atoms be created?

Einstein and his contemporaries proved that atoms can be forced to release their nuclear energies under specific conditions. This tells us that Nature made atoms from nuclear energy under some specific and extreme conditions, most likely at millions of degrees and under millions of pounds of pressure. Such conditions exist in nuclear masses like our Sun and stars.

Scientists have identified a number of types of atoms on the surface of the Sun, from hydrogen to iron. The glowing hot gases of the Sun are made of about the same chemical elements composing the crust and atmosphere of Earth, including carbon and hydrogen, the building blocks for methane. Since these clues from the Sun suggest the strong possibility that Earth's atoms could have been created from nuclear energy in situ, the key question becomes: Does Earth have a nuclear core that created and still creates the atoms comprising its mantle, crust and atmosphere? This question was answered in the affirmative in Chapter II.

With the concept of a nuclear core in each planet, the enigmatic mystery of planetary origins dropped its impregnable veil, and the dawn of reality came sharply into focus. Too good to be true; something must be wrong. But in perfection, where every piece has fitted precisely into the concept, its first flaw yet remains undiscovered after 22 years of piecing it all together and testing it superficially in scientific meetings.

EVIDENCE FOR THE MAKING OF ABIOGENIC METHANE

Fortunately, it is possible to differentiate between biogenic and abiogenic methane by identifying two isotopes of carbon comprising the gas. Biogenic (biological) methane is enriched in ^{12}C, while abiogenic gas consists of ^{13}C (Hoefs, 1980). The high ratios of ^{13}C to ^{12}C in methane, especially deep gas, indicate abiogenic origins. The high ratios in methane from "hot spots" of the Red Sea, Lake Kivu (East Africa), and the East Pacific Rise (McDonald, 1983) suggest abiogenic origins.

Additional background literature by Peyve (1956) and Subbottin (1966) argues that subcrustal abiogenic petroleum migrates up major faults to be trapped in sedimentary basins or dissipated at Earth's surface. Pofir'ev (1974) cites the flanking faults of the Suez, Rhine, Baikal, and Barguzin grabens as examples of such petroleum feeders.

This school of thought gained support in later publications; e.g., *The Deep-Earth-Gas Hypothesis* by Gold and Soter (*Scientific American,* June, 1980). The article presented much evidence that earthquakes and volcanoes release gases from Earth's mantle, and such gases may include methane of a non-biological origin. In a later publication, Gold argues that earthquake outgassing along faults allows methane to escape from the mantle, a process that gives rise to deep gas reservoirs and, via polymerization, to petroleum at shallower levels. Further, the *Reader' Digest* (April 1981) published the article *Bonanza! America Strikes Gas*, which tells of geologists hitting field after field of natural gas deep within the nation's bedrock.

These articles and a number of other discoveries and arguments strongly support both the energy fuels theory and its all-embracing concept of internal nucleosynthesis (first introduced in 1975 as the TIFFE concept: The Internal Formation of Fuels and Elements, later given the I-T-E-M concept label). They add powerful support to the one small, lone voice that argued against the hyssteria of the energy crisis in the 1970s in attempts to convince the establishment of the true nature of the vast reserves of hydrocarbon fuels in Earth's crust worldwide -- a message not heeded until the 1980s.

Today's glutted oil market and reasonable prices attest to the warranted faith in the concept. But even at this writing (1996), the origin of these energy fuels remains erroneously attributed to fossils.

THE RISE AND DECLINE OF THE ABIOGENC FUELS THEORY

The tap root of the energy crisis of the 1970s can be traced back to the original theories of the 1830s and 1920s: gas, petroleum and coal were made from fossils. This misconception proved to be one of the most tragic scientific myths of our time. Its implications of very limited fuel supplies had traumatic effects during and after the 1970s debacle.

During 1973-1976 the price of petroleum tripled, and fuel shortages resulted in long lines at gasoline pumps. Automobile manufacturers, air and freight lines, utility companies and their customers all struggled against the surging tide of rising prices, high inflation and soaring interest rates. Some experts predicted that gasoline prices would climb as high as $5.00 per gallon. OPEC seemed in complete control.

It was a time of genuine, irrational fear that fuel reserves would be depleted within a few years. In April, 1977, President Jimmy Carter expressed his alarm to a national television audience, declaring the moral equivalent of war on the energy situation. By 1979, oil prices had nearly tripled again, with devastating effects on the economy.

Many causes and effects were debated. Two of the primary reasons were attributed to (1) the steadily declining exploration and production of petroleum in the USA between 1955-1973, a decline of nearly 70%, and (2) the pervasive fear that the world's fossil fuel reserves were very limited and were being rapidly exhausted. The heightened impact of low supply and high demand, intensified by depletion woes, was traumatizing to many people. Fear and dire predictions ran rampant, at times bordering on hysteria that was duly captured by the news media: experts spreading doom and gloom -- all based on the false gospel of very limited reserves of fuels made from fossils.

Much of the sanity that survived remained in relative obscurity. For example, a copy of my 1979 book *Undermining the Energy Crisis* was given to Ronald Reagan during his presidential campaign in Louisiana. Its energy fuels concept argued for maximum production of fuels from the vastly under-estimated reserves in the USA, while alleviating the unwarranted fear of rapid depletion of the world's supplies. Further, it predicted that the price of gasoline in the 1980s would stabilize at about $1.00 per gallon whenever supplies met demand.

It is not known for certain that he read the book, but immediately after Reagan's election, the new President issued a three-word order to oil companies: Produce, produce, produce. Consequently, the book's predictions proved deadly accurate. What had made it possible to make such precise short and long term forecasts in opposition to so many prevailing expert opinions? The answer came from understanding how these vast stores of energy fuels were created in Earth's crust -- not from fossils, but via the processes of internal nucleosynthesis and polymerization of these hydrocarbon fuels.

As predicted by the new "energy fuels" theory, and quoting from *Financial World* magazine of November 13, 1990: "...the worldwide glut of oil in the ground, estimated to be over one trillion barrels and rising yearly, will find its way to the market, driving prices inexorably lower." Other estimates run as high as six trillion barrels, and even that figure will prove to be much too conservative.

The latest verification of the prediction made by and since my first publication in 1975 on the origin of unimaginably vast amounts of abiogenic natural gas in Earth's crust appeared in *Science* magazine, 28 June 1991. In his article *Fire and Ice Under the Deep-Sea Floor,* Tim Appenzeller describes ubiquitous gas hydrates occurring naturally in deep-sea sediments and under the Arctic permafrost in staggering amounts. The total amount of this gas worldwide has been estimated as equivalent to some 10,000 billion metric tons of carbon -- twice the carbon of all known reserves of gas, oil and coal.

And who can imagine the vast reserves and the source of the fuel that exists beneath these gas hydrates? Once again, these new estimates of the world's gas reserves will prove woefully inadequate.

The most significant and exciting aspect of these discoveries is the powerful evidence they offer of the manner in which these fuels were made (and are still being made, and will continue to be made throughout future eons, perhaps far longer than mankind will survive on Earth). These and all future findings will fully vindicate the warranted faith in the 1973 concept of internal nucleosynthesis and polymerization of hydrocarbon fuels.

However, the fossil fuels and condensation/planetesimals/accretion beliefs are firmly entrenched in the scientific literature. They will be very difficult to displace. But time and evidence are definitely on the side of the FLPM/IN concept, and it does seem only a question of time.

Interest in the origin of natural gas peaked in the mid 1980s primarily as a result of the annual Spring meeting in 1985 of the American Association for the Advancement of Science (AAAS). A session on origins of fuels featured Professor Thomas Gold's excellent presentation of strong evidence for the abiogenic origin of deep methane gas. Attempts made through AAAS channels to permit presentations of my findings since 1973 to reinforce Gold's findings proved futile, in spite of the fact that several of the program's scheduled speakers failed to show.

Although given its due publicity, the concept was destined to fall out of favor because of two reasons. First, the information I could have contributed in the session would have completed the big picture of the abiogenic origins of all hydrocarbon fuels (gas, petroleum, coal). However, Gold's singular attempts to fit his version of the origin in with prevailing beliefs -- one being that methane came from outer space and was trapped deep inside during Earth's early formation -- were doomed to failure, simply because they are unrealistic, and they simply fail to mesh with the facts. Secondly, through Gold's persuasiveness, a deep well was drilled later in Sweden to test the theory. Since no fuel was found, interest in the abiogenic origin waned considerably.

The concept of the abiogenic origin of methane and the subsequent polymerization to petroleum erroneously became known as 'Gold's theory.'

Until the facts in the situation are sorted out and faith in the concept is restored, perhaps by some future breakthrough, this vital key to the origins of atmospheric, crustal and internal components of all planetary spheres via internal nucleosynthesis now faces an even tougher road to the full credibility it warrants within the scientific community. But this can be done only if scientists are permitted to hear the full, factual story. It is hoped that this book is the first step along that long trail to restored credibility.

Even without having a chance for a full understanding of the now-completed FLPM/IN (formerly I-T-E-M) concept, Gold and many others have added significant corroborative evidence to the new energy fuels theory (EFT) since its conception in 1973. Now structured with an ever-increasing multitude of interlocking, incontrovertible facts, the revolutionary concept does indeed seem destined to displace the more speculative beliefs about the origins of hydro-carbon fuels and other matter comprising our planetary systems.

REVIVAL OF THE ABIOGENIC FUELS THEORY

Perhaps the breakthrough hoped for above has begun. In the same project study at Georgia Tech (referenced above), nickel and a number of other metals were found in close association with the hydrocarbon fuels. In the common knowledge that metal catalysts generally play a prominent role in polymerization processes, one or more of these metals seemed to be involved here. The presence of these metals, helium, other gases and oils usually associated with natural gas deposits is explained by the nucleosynthesis processes within Earth. For obvious reasons, such magmatic mixtures are more the rule than the exception.

Additional strong evidence on the abiogenic origin of hydrocarbon fuels was presented by Frank Mango at the August 1995 meeting of the American Chemical Society. Mango, a research scientist in the geology/geophysics and chemical engineering departments at Houston's Rice University, disclosed that metals such as nickel played a major role in the generation of natural gas.

It all began when the particular light hydrocarbons he was examining "had structures that were fundamentally a contradiction to existing views on the origin of petroleum." Mango's conclusion that "the origin of light [and crude] hydrocarbons could not be the thermal breakdown of biological molecules" is further confirmation of the abiogenic Energy Fuels Theory of 1973.

However, due again to the prevailing belief in the FFT, Mango was thrown off track when he concluded that the nickel promotes "the conversion of decomposing organic debris into natural gas."

Evidence indicates that natural gas forms via the natural affinities of its component elements, carbon and hydrogen, which later polymerize (usually with the aid of metallic catalysts) into ever larger and varied molecules of gas and petroleum (light to crude). Then, in reaching the third and final stage, the crude petroleum, always in the presence of numerous other elements, polymerizes, cross-links and solidifies into coal.

The key point in Mango's finding is the *further confirmation that the origin of hydrocarbon fuels could not be of a biological nature, but are indeed of an abiological nature.* Perhaps this could be the spark that re-ignites the interest of scientists in the true origin of abiogenic fuels: the seed theory of 1973 that eventually grew into the FLPM/IN concept during the next 22 years.

FROM ENERGY TO MATTER TO LIFE: A Reasonable Continuity.

Scientists have much evidence to show that Nature starts with the simpler elements as building blocks to construct ever-larger molecules. Atoms and molecules with affinities for each other arrange themselves properly and assemble together, under proper conditions, into more complex molecules. Nature's self-assembly process has been recognized by biologists as one commonly found throughout biology. Much like a jigsaw puzzle coming to-gether to make the correct picture, the pieces stick together by hydrophobic and electrostatic interactions, according to David S. Lawrence of State University of New York in Buffalo.

Other laboratories have simulated Earth's primitive atmosphere (methane, nitrogen, water vapor) in experiments yielding all five bases that make up the more sophisticated building blocks of the genetic code. Four of these bases (cytosine, guanine, thymine, adenine) form DNA, the double-helix molecule that spells out the instructions for all living things. The fifth critical base, uracil, substitutes for thymine to make RNA, which acts as a master slave to carry out DNA's orders.

The key point here is that life chemicals are formed easily by duplicating Nature's primitive conditions. Thus the process of chemical evolution must have been relatively simple. This lends strong credibility to the EFT theme that all matter is built with basic building blocks (atomic elements) under proper conditions of temperature, pressure and time specific to the matter in question.

Located with an underwater sonar system in 1993, a natural gas deposit was discovered 170 miles east of Charleston, South Carolina. Based on preliminary mapping, USGS scientists estimated that the area could contain more than 1,300 trillion cubic feet of methane gas locked up in hydrate deposits. If extracted, that volume would serve the country's needs for more than 70 years, based on 1989 consumption levels. However, extracting gas from the deposit located more than a mile beneath the ocean surface would not be a simple matter.

The deep-sea gas vent is ringed by an unusual formation of mussels, clams and other mollusks. Apparently, these life forms have adapted to their dark, gaseous environment by deriving energy from the vented methane. This situation illustrates three important points:

1. The continuity in the chain of dependency in the linked cycles of evolution: the energy-to-matter-to-life relationship in which everything is connected to everything else.

2. The environmental survival principle: adapt or die.

3. These findings are indicative of the huge volumes of abiogenic gas already discovered and the vast deposits yet to be discovered in the crust worldwide.

All three factors are vital aspects of the FLPM/IN concept of an inter-meshing continuity of events in which our SS was created in the energy form from which our planets and their planetary systems are evolving through five stages of evolution in full accord with the laws of physics and chemistry.

THE UNIVERSAL LAW OF CREATION OF MATTER

Nature's procedure for making its huge stores of energy fuels can be expressed as follows:

ENERGY to ATOMS to MOLECULES of GAS to OIL to COAL

These processes of Nature are completely reversible by mankind. Beginning with coal, scientists can extract its original petroleum, reduce it to gas molecules, which can be separated into atoms. And as Einstein and the atomic bomb illustrated, atoms can be forced to release the nuclear energy from which they were created.

By expanding this formula to include all matter made from energy, we can derive a simple Universal Law of Creation of Matter (ULCM):

ENERGY to ATOMS to MOLECULES to GAS to LIQUID to SOLID

Planetary cores supply nuclear energy under various ideal conditions of extreme temperatures and pressures for forging atomic elements, the building blocks of matter. These atoms eventually combine and evolve into the countless molecular configurations of Nature's handi-works. To cite a basic

84

example other than methane, the waters of Earth were made by internal nucleo-synthesis of the hydrogen and oxygen atoms that subsequently combined to form the countless molecules of water comprising Earth's oceans and other water systems.

Internal conditions vary from planet to planet, primarily as a function of core size. The large gaseous spheres composed of lighter elements are products of huge, medium-density, open-to-space cores, while rocky planets containing both lighter and heavier elements are products of smaller, higher density, en-capsulated cores. Additionally, distances of planets from the Sun play a role in shaping their surface characteristics; e.g., frozen gases versus liquid surfaces. Thus, planets will differ in composition and outward appearances as functions of size and distance from the Sun.

RESOURCES AND RAMIFICATIONS: A perspective of the 1980s.

Most scientists recognize that matter was and is indeed transformed from energy particles. But it may be some time before they agree that Earth created its own atmosphere and crust with atoms made in its own internal nuclear furnace. And in forming crust, Earth created (and is still creating) its own energy fuels, beginning with atoms of carbon and hydrogen that combine to create methane, then polymerize into higher gases and petroleum, some of which inundated lowlands, then cross-linked and solidified into coals.

Since these events occurred worldwide, it is reasonable to expect that such fuels can be found almost anywhere in the world if one drills deep enough in the right places. And since these reserves of energy fuels appear plentiful, and nuclear fusion sources are on the horizion, it is quite possible that mankind will become extinct long before fuels are exhausted.

In view of the new concept of creation of hydrocarbon fuels, whole new vistas have been opened to mankind. The probability that these energy fuels might not have the limitations of finite "fossil fuels" has worldwide ramifi-cations. When proven more conclusively, the EFT will bring new perspectives to fuel suppliers and to energy dependent industries, while greatly altering the world's economic and political outlooks for many centuries.

The notion that our supply of energy fuels may be nearly limitless is a certain deathblow to the stranglehold exercised by OPEC in the 1970s and to unreasonably high fuel prices during that unwarranted energy crisis.

In the long term view, the most popular sources of energy should be natural gas and nuclear fusion. Coal and petroleum, both less plentiful and more slowly renewable than gas, should not be burned, but should be utilized for more sophisticated purposes: the manufacture of chemicals and the assurance of transportation needs.

Certainly the fear of running out of fuels should never again be a factor in precipitating an energy crisis. The notion of "fossil fuels" is indeed a fossil of antiquated thinking, the relic of the outdated concept of the 1830s.

A SECURE CONCLUSION

This condensed book of the FLPM/IN concept presents a fundamentally sound, factual argument for a revolutionary version of the origin of our Solar System and the evolution of its planets and their planetary systems by means of natural laws. As with a giant jigsaw puzzle in which all pieces interlock precisely in place, it presents a beautiful and complete picture that offers a new perspective in which the enigmatic anomalies of the SS can be understood, and all relevant discoveries of the space probes interpreted more logically.

In the new perspective, scientists will be better able to understand the true cause of planetary quakes (e.g., earthquakes): explosions that cause land movements, rather than the other way around. Certainly, one cannot argue reasonably that other planetary quakes are caused by sudden land movements on those planets and moons.

Subsequently, the source and cause of lightning will be understood. Recently, scientists have recognized that ground-to-cloud lightning is a reality -- actually the norm, rather than a figment of the author's imagination. We can look to Earth's core for the supplies of electrons that make lightning possible. Inevitably, the understanding of all phenomena of Earth and the SS will follow.

But the path will not be easy. Obviously, the strong entrenchment of the prevailing Accretion Disk theory (the dust aggregation/planetesimals/accretion hypothesis) in the scientific literature presents one of history's most formidable barriers to change in direction of scientific thought. However, history teaches that unsatisfactory concepts are replaced eventually by more satisfactory ones.

Inevitably, time will correctly judge the two viewpoints.

CHAPTER IV

MOONS, PLANETARY RINGS AND COMETS

ORIGIN OF OUR MOON

Our Moon is a gleaming silver globe that for centuries has inspired poets, artists, musicians and lovers: a huge, silent barren ball of rock that travels around Earth in slightly less than one month's time. While it appears to be a pale silver in the daytime, it is actually a dark brown color. Brown is the color of cooled lava, pumice (volcanic glass) and igneous rocks that comprise the surface. Jagged, rocky mountains stretch across part of the Moon.

The silver sphere is 2,160 miles in diameter and one-eightieth the mass of Earth. Its surface gravity is only one-sixth as strong as Earth's surface gravity; consequently it is too weak to retain an atmosphere. Any water or light gas vapors formed in the past would have evaporated immediately into space.

With its mountains, extinct volcanos and moonquakes, it is easy to visualize the Moon as simply a smaller Earth without atmosphere or water. One can expect to find most of the same elements comprising both spheres. However, due to different internal conditions effected by size difference, the ratios of created and of retained elements should and do vary considerably.

Between the first landing on the Moon in 1969 and the last one in 1972, some 850 pounds of rock samples were brought back for analyses. The abundance of some of the main elements -- silicon, magnesium, iron, manganese -- in the two spheres matched. Refractory substances such as aluminum, calcium oxide, chromium oxide and titanium that are difficult to vaporize were quite different, as predicted by the IN concept. The Moon samples showed twice as much as Earth's contain. The biggest differences showed up in the more volatile substances, such as potassium and sodium. The Moon has much less than Earth. No water was found on the Moon.

In 1982, Kirk Hansen of the University of Chicago suggested that the changing rotation rate of Earth determines the rate of tidal dissipation over geologic time. His calculations argue favorably for simultaneous creation of Earth and Moon, thereby adding credibility to the contemporaneous geometric birth of the "double-planet" expounded in the FLMN/IN concept as early as 1980.

According to the FLPM/IN (formerly the GBSST) concept, Earth and Moon were formed contemporaneously from the SEM as it sped beyond the Sun. Coupled into a giant dumbbell formation since then, the two masses trace an intertwined pattern in their revolutions around the Sun. The center point of the dumbbell's mass, rather than the center of Earth, moves in a smooth ellipse around the Sun, causing each of the two masses to follow a serpentine path.

The results from the Clementine survey of the Moon yielded the first complete global portrait of Earth's orbiting partner. Among the findings in 1994:

¶ Volcanic activity as recent as a billion years ago.

¶ A wildly variable crust.

¶ The possibility of ice in the shadow of the south pole.

¶ A crater large enough to span the USA from the East Coast to the Rocky Mountains.

¶ A fresh-looking crater that may have been made in the 12th century and recorded by monks.

Scientists had thought that nothing much had happened on the Moon in the last 3 billion years. The discoveries brought the realization that scientists do not understand the Moon as well as they thought they did.

Clementine's 71-day rendezvous with the Moon revealed a topography marked by steep peaks rising to 10 miles higher than the lowest valleys. The deepest valley is rimmed by the highest peaks: a vital clue to the early stages of formation when the forces of isostacy and ejection worked together to push up the huge rim of pliable, amorphous material that solidified as mountains, while the remaining ejecta sank, filling the void below and forming the deep valley.

Evidence of recent lava flows in the Schröedinger basin suggests that the Moon was erupting with volcanos perhaps 2 billion years after it was believed to have settled down. Such volcanos represent the second stage of mountain building in which ejecta must find a way out from beneath the now-solidified crust. Just as on Earth and other planets, virgin ejecta pushes through crustal places of least resistance to its pressurized flow to build the tall volcanic outlets.

This new evidence fits precisely into the FLPM/IN concept in which these surface features can be traced to their source: internal nucleosynthesis. Additional evidence came in 1995 from the McDonald Observatory in Fort

Davis, Texas. Observations made during a total lunar eclipse in 1993 (released in 1995) show the faint glow of sodium gas some 9,000 miles in the atmosphere surrounding the Moon -- an altitude nearly twice as high as previously observed. Both the sodium and the outgassing of thin vapors from surface craters can originate only within the Moon. Here the IN processes appear to be nearing their final stages, as attested by occasional small moonquakes and the weak outgassing from craters.

The one big question left unanswered in many minds by the space probes is the Moon's origin. A growing consensus among astronomers favors the "giant impact" hypothesis in which the Moon may have gotten its start 4½ billion years ago when a planetary projectile about one-seventh Earth's mass collided with our planet. The energy of the collision crushed and vaporized major parts of the two masses, sending out a high-velocity jet of material at temperatures as high as 12,000°F. Within a few hours, some of it came back together far enough away from Earth to remain in orbit. Earth itself re-formed as a combination of the large mass and the bulk of the projectile.

Proponents of this concept claim that it appears to explain the chemical findings from the Apollo mission; e.g., the moon rocks brought back lack water, sodium and other volatile materials -- precisely the materials that would boil away in the rapid vaporization after impact. And some scientists believe the concept explains why veins of gold and platinum lie shallow enough in Earth's crust to be mined!

But not all scientists are satisfied with such a scenario. First, two of the above statements concerning sodium are contradictory. The 1993 observations of the faint glow of sodium gas surrounding the Moon up to an altitude of 9,000 miles directly contradict the 1980s impact scenario in which all sodium boiled away 4½ billion years ago. Further, the potassium and sodium reported in the early analyses of Moon rocks could not have survived such an impact.

And why would gold, platinum and other heavy metals be found only in a projectile and not in the larger mass of Earth? How were these materials made in the projectile? Would all of Earth's other mined materials have been delivered by the missile? Does this impact scenario apply to all other moons of all other planets? Aren't spherical moons all made in the same manner? Did the moons of Jupiter and Saturn result from projectiles bouncing off their clouds? Why didn't the exploding "asteroids" of Comet SL9 bounce off Jupiter to form more moons?

How dependable were the computer simulations that seemed to confirm the hypothesis? Were their results any more dependable than the erroneously predicted results of the powerful computer simulations of the collisions of Comet SL9 with Jupiter? Wouldn't the same type of simulation erroneously show that all moons of all planets were formed in similar impact scenarios? The fallacy of the impact hypothesis becomes obvious under the pointed finger of question.

As one scientist put it, "Books and articles supporting this giant impact hypothesis of lunar origin are more a testament of our ignorance than a statement of our knowledge."

As with the fallacious dinosaur theory, why is it necessary to look to outer space when better answers can be found more readily by looking to the nucleosynthesis processes within planets? In reality, all anomalies relevant to all moons and planets are explainable within the realm of the FLPM/IN concept. For example, the Moon once generated a magnetic field which may have been nearly twice as strong as the present-day magnetic field of Earth, as shown by Runcorn *et al.*, who used magnetized lava rock from the Moon as evidence.

The proof of the declining strength of the Moon's magnetic field strength is powerful evidence that such magnetism is of nuclear energy origin rather than of iron or rocky core origin. The large decline is attributable to the dwindling size of the core as its energy transforms into matter.

Other observations and studies of moons and planets have revealed magnetized lava rocks, the increasing of interior temperatures as a function of depth, the size and density of cores, the equatorial bulges, the convection of materials, the volcanic layers, the chemical composition of crusts, the ejecta and impact craters, the advanced stages of evolution, etc. All are symptomatic of nuclear cores.

In contrast to the impact concept, evidence supporting the FLPM/IN version clearly shows that Earth and Moon formed contemporaneously and both have evolved via internal nucleosynthesis into the rocky fourth stage of evolution. The Moon is nearing the fifth and final stage: an inactive sphere.

In the final analysis, the Moon is nothing but a small planet that evolved alongside Earth and in the same manner as Earth, partners from the beginning.

Shelley expressed it beautifully when he wrote: "...that orbed maiden, with white fire laden, whom mortals call the moon."

THE FAR SIDE OF THE MOON (1992)

USA and Soviet lunar-orbiting craft have photographed portions of the far side of the Moon. Painstaking analysis of computer tapes indicate that some of the images provide an accurate, multi-wavelength portrait of parts of the far side. The images from Mariner 10 and Galileo, each depicting a different part of the far side, are helping astronomers decipher in detail the composition of the Moon's hidden half. Surface composition provides crucial clues to the nature of volcanic eruptions and other upheavals that shape the surfaces of spheres.

Scientists have already mapped much of the chemical composition of the Moon's near side. Analysis of the lunar rock samples brought back by U.S. astronauts and unmanned Russian craft have proven extremely helpful.

Two types of terrain form the Moon's outer surface. The light-colored highlands represent the brighter, lower density minerals. In contrast, the dark plains are regions identified by Galileo and his Renaissance colleagues as maria, a Latin word meaning *oceans*. To them, the dark plains appeared to be bodies of water. Researchers generally agree maria are the results of volcanic eruptions: lava flows from 4 billion to 2.5 billion years ago.

Images from the spacecraft confirm that the maria are much less abundant on the far side than on the near side of the Moon. This suggests earlier and then milder volcanic activity on the far side -- the latter due to the earlier, thicker crust there. But why would the crust be thicker on the far side? And when did the maria form?

Evidence suggests that early volcanic activity that formed maria was extensive; some maria existed more than 4 billion years ago. According to the IN concept, the Moon's crust formed in the same manner as all other spheres of the SS, passing through the common stages of evolution while cooling from hot energy to cold matter.

During the transition stage, the Moon's thin crust struggled to survive and grow, eventually encapsulating the sphere with a rough, flexible blanket. Ever thickening, the soft crust became a thick, bubbling, mud-like caldron, stirred by escaping gases, matter and fireballs, all newly created within. These escape outlets hardened into circular patterns now identified as impact craters.

Meanwhile, the heavy atmosphere thinned and eventually vanished. Volcanic flows became common, covering many of the craters, even as others were being formed. Volcanic mountains and lava plains formed from huge outpourings of virgin materials that found the paths of least resistance to their upward pressures. Fireballs shot violently from within and fell back to the surface, sculpting the landscape with a multitude of ejecta and impact craters into the circular patterns and dark maria visible today.

Instruments that view the surface at several wavelengths can detect the dark, underlying maria, called crytomaria, mixed with the lighter-colored shroud of highland crust.

Patrick Moore, in his book *New Guide to the Moon* (1976 edition), presents a strong case for the formation of the craters of the Moon by internal forces rather than by impacting meteorites. Included is convincing evidence by Allan Mills, who successfully produced model craters strikingly similar to those of the Moon. He accomplished this proof by introducing a pressurized stream of gas below a particular material, which behaved as if liquid while the bed expanded. Bubbles appeared, venting through the muddy material to produce the crater-like appearance of the Moon.

Analysis of views of the far side show that the Moon's hidden half contains far fewer maria than the near side, and far lower concentrations of titanium oxide than do many maria on the Moon's near side. This evidence and the analysis of moonquakes and human-generated disturbances on the Moon have revealed that the lunar far side has a much thicker crust than the near side. To the obvious question of why, the IN concept offers a reasonable answer.

As contemporary partners for some five billion years, Earth and Moon evolved with the near side of the Moon always facing Earth. During their evolution from energy masses to rocky spheres, the lunar far side remained exposed to the severe cold of space, while the near side basked in the heat from Earth's nuclear mass, a small secondary sun. These extreme conditions, along with the slightly larger centrifugal force acting on the far side, caused the crust there to evolve quicker and thicker than the much warmer near side evolved. Under these extremely different conditions, the two sides naturally evolved differently.

Noticeably more craters and titanium oxide on the near side are indicative of a later, slower, longer stage of ejecta cratering. The difference in the amounts of titanium oxide is attributed to the fact that larger volumes of heavier elements such as titanium are produced in the later stages of evolution of any sphere. The steady increases in internal temperature and pressure -- caused by the huge energy-to-matter expansions within -- as the sphere become more tightly encapsulated accounts for the larger amounts of titanium oxide produced at a later time and over a longer period of time on the near side than on the far side of the Moon. The thinner crust adds another advantage to the near side by providing an easier upward route for the chemical ejecta.

One interesting observation here: The titanium produced on the Moon is analogous to the tellurium produced on Venus, the iridium produced on Earth, the magnesium produced on Neptune, and the abundant metals observed at the impact site of the G fragment of Comet SL9 on Jupiter -- all are in situ ejecta products of nucleosynthesis.

In summary, the differences in the number of craters, the thickness of the two crusts and the amounts of titanium oxide are indicative of the differences in the rates of cool-down and solidification of the crust on each side of the Moon. These different rates, in turn, are functions of the outside temperatures to which the two sides were exposed and to the steady increases in internal temperature and pressure as encapsulation progresses as a function of time.

Other questions challenge the impact craters hypothesis. If by impact, why such a noticeable difference in the number of craters on the two sides of the Moon? And why so many craters altogether on such a tiny target? Why are all craters lined up precisely with the center point of the spherical Moon? Why don't any of the craters show a skewed configuration typical of a near miss projectile hitting the surface at a sharp angle?

In conclusion, the evidence supporting the FLPM/IN concept argues strongly against the giant impact hypothesis of the Moon's origin, while presenting a valid case for the origin of Earth and Moon as contemporary partners during their simultaneous evolution via nucleosynthesis from energy masses to rocky spheres.

Although beginning their evolution simultaneously, Earth and Moon have evolved at different rates because of the size factor. The Moon's energy core is nearing depletion, as attested by very weak signals from within. Approximately two-thirds of Earth's core has been depleted in nearly 5 billion years, leaving an estimated 2.5 billion years until it become inactive.

Meanwhile, its orbit will have moved approximately 20% closer to the Sun. During this long interval, Earth's climate will cycle through many changes while growing ever more barren. Our planet will follow Mercury, Moon, Pluto, Mars and Venus in becoming inactive spheres. Expectations that humankind can survive the gradual drift into these drastic changes seem unreasonable.

HOW PLANETARY RINGS WERE FORMED

In 1610 Galileo first observed Saturn's peculiar form -- recognized several years later by Christian Huygens as rings. The next set of rings were detected in 1977 by James Elliot and his colleagues at Cornell University when they noticed that the light from a star blinked several times just before Uranus occulted it. Voyager 1 sailed past Jupiter in 1979 and Saturn in 1980. Voyager 2 visited Jupiter, Saturn, Uranus, finally sailing past Neptune in 1989. The spacecraft made the first detailed images to confirm the rings, and discovered rings around Jupiter and Neptune during the extensive SS tour.

How can the formation of these rings be explained? By observation, the rings in each case are located at the 'equator', the circumference of maximum centrifugal force. In the perspective of the IN concept, substances (gas, liquid, solid) were forcefully ejected from these spheres and collected at the circumferences of maximum force, and remained suspended in orbits around each active planet. The greater the escape velocity/mass ratio, the higher the orbit of each substance. Regardless of any change in the tilt of the planetary axis, the rings will remain firmly in equatorial orbits while their velocities decrease over billions of years, eventually to drop onto the planet to become parts of the crust.

The same principles can be applied to other satellites of planets, whether Nature-made moons or man-made craft, provided the orbiting masses remain below a critical mass/velocity/gravity/distance relationship. Any mass exceeding this yet-to-be calculated relationship will gradually drift away from its central body. Uranus, Neptune and our Moon are examples of orbiting bodies known to be undergoing such drifting.

To some degree, when blasting satellites into orbit, mankind replicates Nature's procedure for orbiting the materials that formed the planetary rings and some of the smaller moons. In this perspective, the discoveries of the rings of Jupiter, Uranus and Neptune were predicted by the IN concept before the author learned of their existence. By the same logic, it can be predicted that Pluto, because of its small size, is well beyond this ring stage of evolution.

Why rings on the larger outer planets and not on the smaller inner planets? The reasons can be found in their sizes, which control the relative rates of planetary evolution. In the gaseous stages of their transformations, all planets had rings of ejecta: gases, solids and liquids that quickly froze solid. Like a flower in full bloom, Saturn is at the zenith of planetary rings, the goal that Jupiter will attain some time in the distant future -- the stage through which all small planets of our SS already have progressed.

The smaller the planet, the earlier it passed through the ringed stage of its evolution, which ended when all materials had fallen out of their orbits in full accord with the natural laws of gravity. Nothing can stay in orbit forever. Uranus and Neptune are in the twilight time of passing out of their ringed stage, while Pluto should have passed out of it a long time ago.

Ejecta created via internal nucleosynthesis is the key phrase that firmly ties together all things: planetary rings, craters, volcanos, planetquakes, moonquakes, fireballs, comets, meteorites, the moons of Mars (Phobos and Deimos), lightning, electromagnetism and all other phenomena involved in the evolution of physical worlds. All are interconnected; all can be traced to a common origin: the internal transition of energy to matter (ITEM) in each sphere. In this perspective, all past and future discoveries of the space probes will be understood.

JUPITER'S RINGS: A Giant Leap Forward

Until early 1992, the prevailing Accretion Disk theory (based on the condensation/planetesimals/accretion hypothesis) taught that planetary rings consist of leftover debris that did not accrete. In February of that year, the news media announced that the Ulysses spacecraft boomeranged past Jupiter on Saturday, flying through *intense radiation and an orbiting ring of volcanic debris* on its way to study the Sun.

The underlined words above represent a giant leap forward in the right direction of thought in the scientific community concerning the origin of planetary rings. The *nuclear energy core of Jupiter accounts for both the volcanic debris and the intense radiation* experienced by Ulysses. Both findings bring current beliefs about planetary origins sharply into question. Since both findings are products of internal nucleosynthesis and were predicted

95

by the IN concept, they are powerful evidence supporting the FLPM/IN concept that foretold these discoveries as far back as 1973.

COMET HALLEY AND ITS LAST FAREWELL

In March, 1986, five space probes encountered Halley's comet to photograph its nucleus and to analyze its composition during its most recent periodic visit to the Sun. The photographs made by Giotto, the closest probe, revealed some real surprises. The pictures and data beamed to Earth showed a velvet black, peanut-shaped nucleus with a surface full of pits. According to the IN concept, the black surface is the carbonized crust encapsulating its nuclear energy mass, and the surface pits are ejecta holes or friction marks formed in the past by powerful jets of forcibly ejected materials.

Vega reported evidence of water, carbon dioxide and hydrocarbon molecules among its findings, all products of combustion. Temperatures within the comet's thin plasma coma were recorded to be well above the boiling point of water. *Computer enhanced color images were identical to those of the background stars, indicating similar high temperature gradients throughout each of the photographed bodies.* These high temperatures received little notice and less publicity, but remain a significant factor in deciphering the true nature of comets. In spite of this contradictory evidence, many scientists still cling to the belief that comets are rocky snowballs!

Violent, bright jets of gas and fine matter spew profusely from vents (or craters) in the nucleus. The jets appear to be firing out through two or three vents comprising only a small fraction -- about 10 percent -- of the total surface, leaving some 90 percent of the surface inactive. This fact of no losses from the black surface is highly significant: It means that materials forming the tails and the coma (with a radius of 600,000 miles) could have come only from within the nucleus: the source of its creation.

Dimensions of the nucleus were estimated as 16x8x8 kilometers. Its surface area is approximately four times larger than expected, and thus, the low albedo of 2 to 4 percent means that Halley's comet is the darkest of all known bodies in the SS.

Scientists have difficulty in reaching firm conclusions about the nature of the observed jet-like features, and in interpreting new information to fit the concept of a dirty snowball of pristine materials that supposedly formed the SS. The basic reason for these difficulties lies in the fallacy of the antiquated accretion disk hypothesis that evolved from the Kant-LaPlace hypothesis of a gaseous dust-cloud origin of the SS -- a tragic myth that continues to lead many scientists down blind alleys in which nothing fits together.

The Giotto pictures confirm that inert shells do develop on cometary surfaces, much like crust forms on planets -- an exciting quantum leap. The next step is for skeptics to realize this startling fact. From there, it will be much simpler to understand how moons, planets and their systems were made in compliance with the laws of thermodynamics.

Both Giotto and Vega spacecraft carried aboard them sensitive instruments to identify gaseous molecules in Comet Halley's coma. They found a veritable stew of molecular fragments known as free radicals, confirming conclusions made previously from telescopic observations. Some 75-80 percent of the radicals pertained to water, a ratio common to Earth's water:land ratio produced via nucleosynthesis.

The remaining free radicals suggest that carbon dioxide, ammonia, methane and hydrogen cyanide would be the end products if these highly reactive radicals could get close enough together to interact and combine. However, most remain too far apart to combine into these compounds. It is no coincidence that the same compositional radicals are found in stars, and the same end products they form are found on all planets and moons of the SS. All were and are produced via nucleosynthesis within the fiery core of each evolving mass.

Other than these gases, and not surprisingly, a major discovery was the existence of tiny, solid CHON particles. CHON is an acronym for the chemical symbols of the compositional elements: carbon, hydrogen, oxygen and nitrogen. While these apparently organic compounds are not biological in origin, they are indicative of the presence of organic compounds that might appear as black soot made via nucleosynthesis within the fiery comet. (Note the similarity with the nucleosynthesis of hydrocarbon fuels in Chapter III). This accounts for the velvet-black surface of the nucleus, perhaps a carbonized insulator against the tremendous heat of its core.

Large quantities of very fine particles were gathered by the dust collectors aboard Giotto and Vega. Chemically, these motes consist of many of the same elements comprising terrestrial rocks: iron, magnesium, silicon, oxygen, etc.

The presence of these virgin elements, created via nucleosynthesis within the comet, further validates the FLPM/IN concept.

Since 1987, observers have detected methanol in seven comets. It now appears that this compound ranks third in abundance among cometary ejecta, behind water and carbon monoxide. The methanol can amount to 5 percent or more (relative to water) of a comet's volatile matter.

The results from Halley's Comet reveal its true nature. Rather than 'dirty snowballs', comets are what they seem: nuclear fireballs ejected from larger nuclear masses (much like flares ejected from the Sun), and powered in their orbits like nuclear-fueled jet engines. This mental picture was brought sharply into focus on the TV program *Science Frontiers* on June 7, 1994. Some eight years after the big event, the public was privileged to watch video close-ups of Halley's Comet zipping through the sky, looking precisely like a jet-propelled rocket in stabilized flight (see illustration).

Records show that the first recorded sighting of Halley's comet was 240 B.C. It has returned periodically every 76 years, each time diminished in size as energy transformed continually into ejected ionized matter. The original gigantic comet must have been a terrifying sight in its early years. In 1456, it frightened everyone in Europe so much that Christian churches added a prayer to be saved from "the Devil, the Turk and the comet." As late as 1910, it still furnished a good, but smaller show. But in 1986, it was a disappointment to observers, who had to look closely to find it.

Then in February, 1991, when Halley's comet was 1.3 billion miles from the Sun, about midway between Saturn and Uranus, it suddenly expanded to 180,000 miles across and shone to more than 1,000 times brighter than normal. The event was startling and unique, totally unexpected so far from the Sun, and remains unexplainable in the perspective of the 'dirty snowball' concept.

However, viewed in the FLPM/IN concept, the spectacular event is understandable: the comet simply exploded in its death throes. It was the last sighting of the famous comet -- the last farewell -- never to return on another mission to brighten Earth's skies. Other known comets have had catastrophic endings; some have split into two or more pieces, others have crashed into larger spheres, including the Sun. The Biela comet broke in two in 1846 and has since disappeared. In reality, each fiery comet exists for a relatively short time, usually measured in centuries, before becoming extinct. Rather than primodial remnants (dirty snowballs) left over from the formation of planets, they are more like a fiery species that refuses to become extinct.

COMETS, ASTEROIDS AND METEORITES: A Close Relationship

One need not look outside our SS to decipher the sources and history of comets and asteroids. The ejecta-cratered bodies of comets and asteroids attest to the fiery history of evolution from hot to cold masses that often end up as burned-out shells like the moons of Mars: Phobos and Deimos. And sometimes we see asteroids like Ida and Gaspra, cratered via outgassing as the hot cometary masses cooled inward after quickly forming a dark shell -- exactly like lava rocks cool on Earth.

Throughout the cycles of formation of crust on planets and moons, various sizes of small to moderate energy masses were forcibly ejected as cometary fireballs from the large masses. Many escaped the gravity of the planet/moon as comets; those that did not escape went into orbits as volcanic debris or simply fell back onto the crust to explode. In most cases, the creation of big holes occurred, some as ejecta craters, others as impact craters.

The article entitled *Rocky Relics (Science News,* 5 Feb 1994) discusses near-Earth asteroids (NEAs) and the possible relationship of comets, asteroids and meteorites. It states that in 1992, the asteroid known as 1979VA was identified as the same object that had been identified in 1949 as a comet (Wilson-Harrington) with a "faint, but definite tail...Researchers now say that the comet and the asteroid are one and the same." This finding adds much confirmatory evidence and credibility to the same conclusion reached by the IN concept in the 1970s. It represents another giant step in the right direction.

Another quote from the same article states that "*several bodies classified as asteroids may once have been comets.* One candidate is the asteroid 3200 Phaethon. It follows the path of small bodies that produce the Geminid meteor shower, the flashes of light visible each December. Meteor showers are typically associated with comets. Tracks of dust expelled by active comets as they pass near Earth's orbit produce the flashes as the dust burns up in the atmosphere. Following this line of reasoning, Phaethon may be a comet masquerading in its old age as an asteroid."

In 1802 Heinrich Olbers, Germany's leading expert on comets, recognized the asteroids orbiting in the fifth orbit from the Sun as fragments of a planet that had disintegrated, exploded. This accurate interpretation accounts for the more than 30,000 asteroid fragments, most of which would have begun their existence as small, fiery comets and smaller, shattered fragments we identify as meteorites: lumps of stone or metal that survived fiery passage through Earth's atmosphere. In all probability, the surviving asteroidal fragments represent only a fraction of the original number of shattered pieces in orbit. Evidence of current orbital patterns indicates the probability that three planets existed in three distinct orbits before the disintegration.

The Phi geometry in Chapter I shows that the Asteroids planet went into orbit as a single mass at 4.24 AU. Later, due to its vulnerable position between the Sun and the giant Jupiter, it broke into three fiery masses that settled near their current orbits (2.2 AU, 3.2 AU, 3.9 AU) before the explosive disintegration of each of the three potential planets resulted in the three wide belts of orbiting asteroids.

While arguing strongly against the current beliefs of many scientists that planets formed from accreted rocky snowballs or mudballs, these findings clearly show the close relationship among fiery comets, cratered asteroids, small moons and meteorites, thereby adding more credibility to the FLPM/IN concept of our planetary origins.

Color-enhanced Halley's Comet showing the concentric spherical layers of enveloping matter indicative of temperature differentials. Photos of nuclear stars in the background showed identically-colored layers and patterns.

What happens to comets that do not meet catastrophic deaths before all their internal energies become depleted? We can look to Mars for two good examples of burned-out cometary masses: its two moons, Phobos and Deimos. Both of the small, irregular bodies exhibit typical characteristics of burnt-out comets: ejecta craters and parallel striations or grooves. The densities of both moons, like the volcanic rocks of Earth, are much lower than that of Mars.

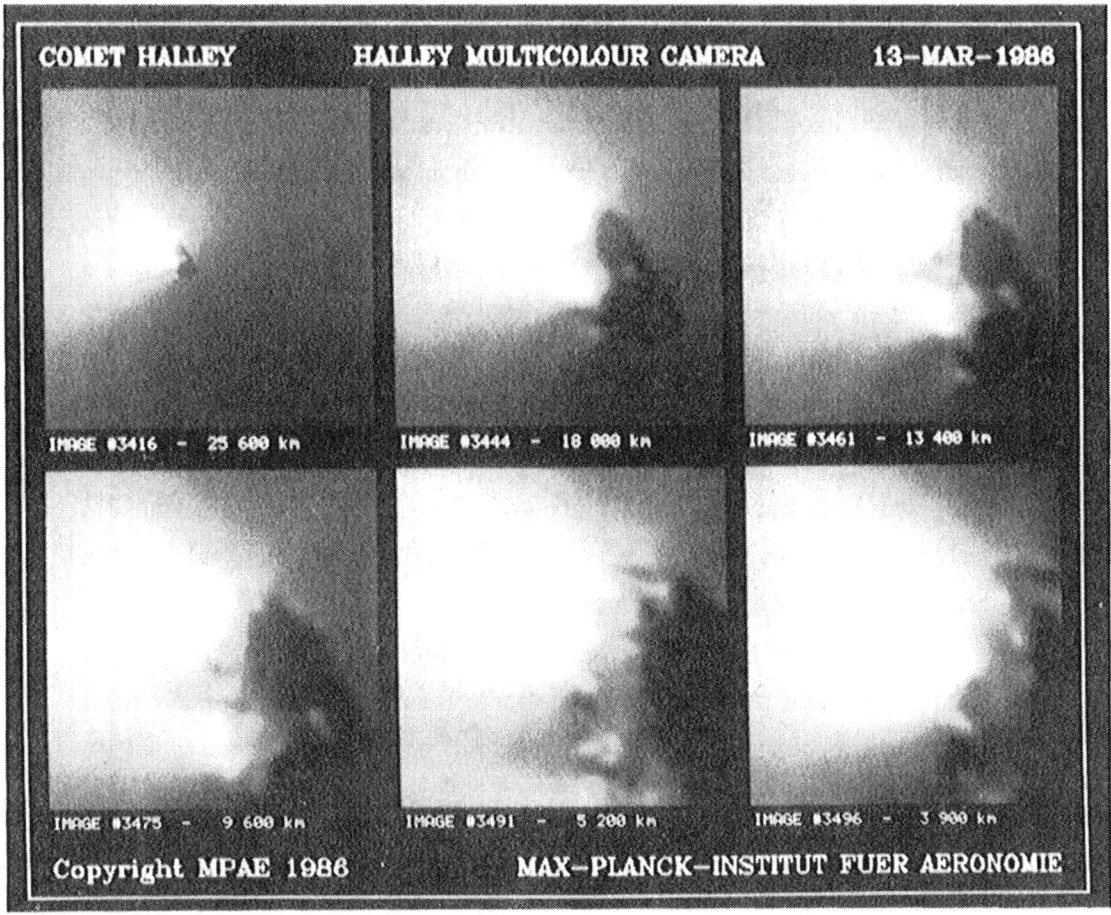

COMET HALLEY HALLEY MULTICOLOUR CAMERA 13-MAR-1986

IMAGE #3416 - 25 600 km IMAGE #3444 - 18 000 km IMAGE #3461 - 13 400 km

IMAGE #3475 - 9 600 km IMAGE #3491 - 5 200 km IMAGE #3496 - 3 900 km

Copyright MPAE 1986   MAX-PLANCK-INSTITUT FUER AERONOMIE

Photos Max-Planck-Institute Fuer Aeronomie, Lindau/Harz, FRG, taken by the Halley Multicolour Camera on board ESA's Giotto spacecraft Interpretation is by the author

These exciting photos of Halley's Comet clearly reveal its fiery nature Note the brilliant white jets firing from the rear of the fireball's irregular crust, locking the nucleus in a precise flight position in each picture -- much like a jet-powered rocket. When the core's nuclear energy finally is exhausted via conversion into matter, the nucleus may remain as a pitted, ejecta-cratered chunk resembling the two moons of Mars: Phobos and Deimos.
 Note Time elapse for photos was approximately one hour

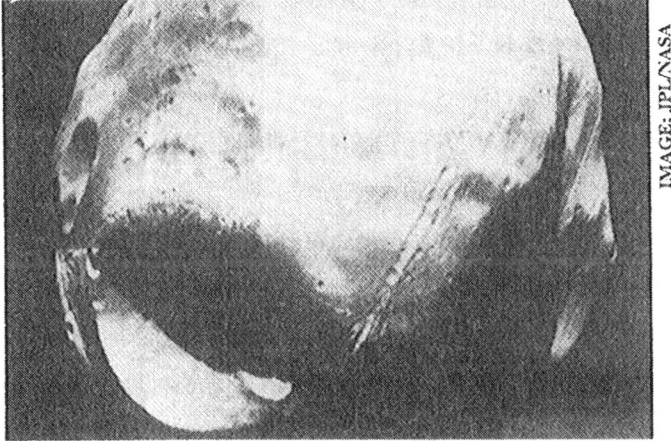

IMAGE: JPL/NASA

Phobos, a burned-out cometary nucleus.

From certain angles, Phobos even bears a striking resemblance to a jet engine.

In 1988, scientists detected traces of sodium and potassium as components of the Moon's extremely thin atmosphere. In 1991, researchers discovered that the atmosphere stretches out into a long tail about 21,000 kilometers long and pointing away from the Sun. Also, they found a corona of atmospheric atoms extending 7,000 kilometers above the lunar surface -- much like the head of a comet. Their observations suggest that the Moon's atmosphere qualitatively resembles that of a comet, featuring a bright coma and an extended, but extremely tenuous tail.

Some astronomers now concede that the Moon has the appearance of a comet. This is a giant confirmatory step toward the realization of the manner in which all SS spheres were made, while showing the close, fiery relationship among comets, asteroids, moons and planets.

These similarities between the Moon and a comet are highly significant and typical of universal objects. They clearly illustrate how both large and small bodies are being created in identical manner via the internal transition of energy to matter (nucleosynthesis) by means of natural laws. In future years, scientists will discover that all active bodies will continue to exhibit these characteristics until their nuclear cores are exhausted.

The first of three deep-space missions to be flown by the year 2000 will feature a 1998 launch of a small spacecraft destined for a flyby of an asteroid and a comet. The 200-pound spacecraft's science instrument payload will include a miniaturized imaging spectrometer that will record the chemical nature of the two targets. The mission should be able to determine the true nature of comets and asteroids, confirm their close relationship, and finally melt away the dirty snowball hypothesis of our planetary origins.

HALE-BOPP: The Next Great Comet?

On July 23, 1995, two amateur astronomers independently discovered their namesake comet, Hale-Bopp, blazing away beyond the orbit of Jupiter 25,000 times brighter than Comet Halley did when it was at the same distance from the Sun. As it swings ever closer, the new Hale-Bopp mass should put on a spectacular show in March and April 1997, a spectacle reminiscent of the early appearances of Comet Halley between 12 B.C. and 1682 (and to an ever lesser

precisely what happens when a huge amount of energy is instantaneously added to a system. The "extra electrons" came from the exploding nuclear fragment. Any dust, if present, would have played an insignificant role. Only a nuclear explosion could have caused the 7-minute "rat-a-tat" static.

Beginning July 16, astronomers were stunned by the tremendous explosions caused by the fragments of Comet SL9 crashing against the thin cloud tops of Jupiter. Each powerful collision far exceeded supercomputer calculations. Comparisons of expectations with actual results were startling, even shocking to scientists. The *Astronomy* magazine (Nov 1994) published a well-written article evoking full agreement that pre-crash expectations were among the casualties. The spectacular explosions were indeed vastly more brilliant and powerful than predicted.

The information in this paragraph is from the record of predictions made by researchers (*Eos*, 5 Jul 1994). "Prior to the big event, three dimensional simulations of the impacts on Jupiter were performed on the 1840-node Ontel Paragon, the world's most powerful parallel computer, located at Sandia National Laboratories, Albuquerque, NM. The key results of the fireball simulation for a 3-kilometer fragment impact were moderate, but seemed sufficient for the event. At 70 seconds after impact, a spherical shock wave has reached a diameter of 700 km and an altitude of 900 km above the clouds. Hidden within the spherical shock wave is the fireball itself, which is a rapidly rising cloud of cometary debris and Jovian atmosphere at a very high temperature (1700 K). Its apparent magnitude of 2 is about one-fiftieth as bright as Jupiter itself. The visible fireballs will be orange, fading to red, and most of the energy will be radiated in the near infrared."

Adding to the predictions, an article in *Science News* (9 Jul 1994) headlined *The 200,000-Megaton Meeting* stated that each chunk is packing a punch that may exceed 200,000 megatons of TNT, and the expanding fireball may be as wide as 100 km when it emerges through the cloud tops within a minute after collision.

These recorded predictions later could be classified as puny when compared with the actual show put on by the impacts. The collisions were stunningly spectacular, the greatest ever witnessed in recorded history. Instead of small, orange flashes one fiftieth as bright as Jupiter, the huge, white-hot explosive flashes were up to fifty times brighter than Jupiter. Expended energy estimates shot upwards to the 250 million megatons of TNT, then to 600 million megatons, creating temperatures of more than 30,000°C.

Fragment G alone propelled a fireball thousands of kilometers above Jupiter's stratosphere and is thought to have yielded at least 6 million megatons of energy. (A megaton is the equivalent of a million tons of TNT). To expend an equal amount of energy, a Hiroshima-type bomb would have to be detonated every second for ten years.

Infrared radiation (heat) from the explosion was so great that detectors at the Keck Observatory were overwhelmed, or saturated. One collision was estimated to be the equivalent of 500 million atomic bombs. Astronomers reported seeing fireballs (from two of the smaller fragments) that erupted up to 1000 km above the cloud tops. After fragment C hit, astronomers at the Keck Telescope in Hawaii took infrared photos. The views show two glowing ovals, each about the diameter of Earth, left by fragments A and C.

Other computers proved equally wrong. First, the fragments exploded at the cloud tops (a crucial clue to their true nature) rather than penetrating deep enough to dredge up Jovian material, disappointing many who hoped the impacts would reach Jupiter's water-rich lower atmosphere. Researchers reported that infrared spectra from Galileo suggest that comet fragment G exploded high in Jupiter's atmosphere, never penetrating the uppermost cloud layer. Analysis of much weaker spectra from the R fragment supports this conclusion. Another team of researchers deduced from absorption lines in their spectra that relatively little atmosphere resided above the G fireball, indicating that the fragment had not plowed deeply into Jupiter.

The anticipated problem of separately identifying Jupiter's water and the supposedly icy water of the comet never materialized. After the largest collision by fragment G, astronomers were puzzled by their failure to find the chemical signature of water in the clouds created by the comet's impacts on Jupiter. "It's puzzling, but we will continue to look for water," stated one astronomer.

Scientists studying one of the early impacts found evidence of two chemical species never seen before on Jupiter: methylene (a carbon compound) and the hydroxyl ion. Both species would seem to require high heat to be formed.

Together, these findings raise serious doubts about prevailing beliefs concerning the nature of these brilliant streakers across the sky. If they really are

COMET S-L 9 COLLISION WITH JUPITER.

A stunningly spectacular explosion of one fragment of Comet SL-9 on Jupiter in July 1994 that far exceeded all predictions. Fragment G alone propelled a fireball thousands of kilometers above Jupiter's stratosphere and is thought to have yielded at least 6 million megatons of energy. A megaton is the equivalent of a million tons of TNT. To expend an equal amount of energy, a Hiroshima-type bomb would have to be detonated every second for ten years. One collision was estimated to be the equivalent of 500 million atomic bombs. The fragments exploded at the cloud tops, another crucial clue to their true nature: nuclear energy fireballs. The picture shows the actual explosion of a nuclear fireball rather than of a dirty rocky snowball. This information is devastating to the Big Bang theory.

dirty snowballs', there should have been obvious signs of copious amounts of water released in the collisions. And how will scientists explain the gigantic discrepancies between the conservative supercomputer predictions and the unexpected stunning immensities of the collisions?

There can be only one reason for these huge discrepancies. Computers will put out correct information *only* when correct information is fed into them. Had they been fed information based on the correct assumption that the fragments were of a nuclear energy nature rather than of an ice-rock composition, the computer responses would have been precisely in line with the stunning results.

The obvious answers to all questions concerning these memorable collisions lead inexorably to the conclusion that the spectacular explosions were indeed of a nuclear energy nature. This rare event should be the final nail in the coffin of the 'dirty snowball' hypothesis, a false speculation that too long has misled scientists down blind alleys.

In reality, comets flashing through outer space are precisely what they appear to be: masses of hot nuclear energy that should not to be confused with inactive asteroids composed of rocky matter that does not burn brightly in their flights through the vacuum of space, thereby remaining invisible to the naked eye.

This new knowledge about the true nature of comets should dispel the notion of 'dirty snowballs' from which our planets supposedly came into being. If not dirty snowballs, then the building material for planets, moons and systems universal must be the one thing that is distributed throughout the Universe, the one thing that comprises all atoms and molecules of matter, the one thing that furnishes a solid basis for understanding all anomalies of our SS, the one thing capable of explaining all the stunning results of the SL9 impacts on Jupiter: nuclear energy. The picture below clearly shows the white-hot nuclear power of nine Comet SL9 fragments on their way to colliding with Jupiter.

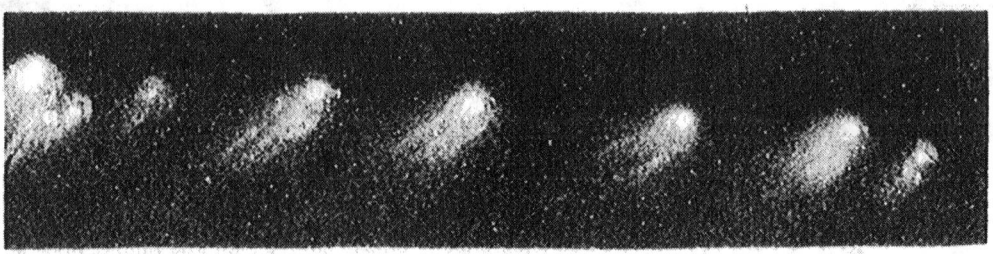

CHAPTER V
A BIG BANG OR LITTLE BANGS?

THE THREE KEY OBSERVATIONS

Three observations provide the fundamental basis for the standard cosmology featuring the Big Bang (BB) theory:

1. The observed expansion of the Universe (usually interpreted in the framework of general relativity as an expansion of the metric of space).

2. The 2.726 K cosmic background radiation (CBR), interpreted as a remnant of the BB.

3. The apparently successful explanation of the relative abundance of the light elements.

In reality, the same observations serve equally well as the fundamental basis for a different concept: the Little Bangs (LB) theory that intermeshes precisely with the geometric origin of the SS and the evolution of planets via nucleosynthesis. Pieced together during the past 22 years (1973-1995), this revolutionary perspective offers answers to the many questions that pose challenges to the BB theory. The two concepts need to be examined closely with wide open minds.

QUESTIONS ABOUT AN EXPANDING UNIVERSE

Hertherington's *Encyclopedia of Cosmology* summarizes the status of the BB theory: "...problems have been numerous and not all of them are solved to the satisfaction of all cosmologists. In particular, large-scale inhomogeneities observed in the 1980s seem to indicate a structured Universe which may contradict one of the foundations of the BB cosmology, the uniformity postulate (or cosmological principle). This and other problems have recently caused some cosmologists to declare the BB theory in a state of crisis. However, since no plausible alternative exists, the almost universal belief in the BB model has not been seriously shattered."

The reason for the fallacy of the last statement is best summarized in the article *Why Only One Big Bang? (Scientific American* Feb 1992) by Geoffrey Burbidge, a professor of physics at the University of California, San Diego. To quote, "Those of us who have been around long enough know that peer review and the refereeing of papers have become a form of censorship. It is extraordinarily difficult to get financial support or viewing time on a telescope unless one writes a proposal that follows the party line. A few years back Halton C. Arp was denied telescope time at Mount Wilson and Palomar observatories because his observing program had found and continued to find evidence contrary to standard cosmology. Unorthodox papers often are denied publication for years or are blocked by referees [Amen]. The same attitude applies to academic positions. I would wager that no young researcher would be willing to jeopardize his or her scientific career by writing an essay such as this."

"This situation is particularly worrisome because there are good reasons to think the big bang model is seriously flawed. One sign that something is amiss is the time-scale problem. The most favored version of the big bang model yields a universe that is between 7 and 13 billion years old. The large range of possible ages derives from uncertainty regarding the rate at which the universe is expanding, a value known as the Hubble constant."

"Comparisons between observation and calculations of stellar evolution imply that the oldest known stars are 13 to 15 billion years old, with an uncertainty of plus or minus 20 percent. The estimated age of the elements in the solar system, based on measurements of heavy radioactive elements, is about 15 billion years, again including some uncertainty. If one accepts a high value of the Hubble constant, and hence a low age for the universe, the simplest big bang model clearly fails, because the universe cannot be younger than the objects it contains. If one chooses a low value for the Hubble constant, it is touch and go."

Recently, the Hubble Space Telescope (HST) provided new data used in revising the Hubble constant to 81-85 km/sec/Mpc. This figure proved disconcerting to advocates of the BB, because any result over 50 km/sec/Mpc reveals a serious flaw in the standard model.

Further, there is an inescapable problem of creation in both the BB and the steady state cosmologies with no scientific solution. How did the singularity or primeval atom originate? Could all matter of the Universe really have been packaged in such a small unit and at trillions of trillions of degrees temperature? If an explosive expansion was nearly instantaneous, why is the Universe now expanding at merely the speed of light? Why wasn't gravity an impeding factor during the time of instantaneous expansion? With the Universe expanding at or near the speed of light (as extrapolation of the factors of the Hubble constant indicate), why are extrapolations backward to the very beginning done _on this basis instead of on the basis of instantaneous expansion_ (an important basis of the BB theory)? Hasn't the Universe always expanded at the speed of light?

Quoting Burbidge again, "Rather than consider alternatives to the big bang, cosmologists contort themselves and propose that the rate of expansion is just small enough to accommodate the oldest well-documented stellar ages. Or they vary the big bang model by invoking an arbitrary parameter called the cosmological constant. In this version of the story, the initial big bang was followed by a waiting period and then a further expansion."

THE COSMIC MICROWAVE BACKGROUND: The 2.7 K Radiation

The second fundamental basis of the BB theory deals with the 2.7 K radiation identified as a relic of the big event. While this interpretation makes a good story, it does raise serious questions that leave the door open to other possibilities. But first, a little background would be helpful.

In 1965, two young scientists, Arno Penzias and Robert Wilson, decided to use sensitive microwave antenna in radio astronomy. Much to their chagrin, they discovered an irremovable background noise in the antenna. Frustrated, they sought help at Princeton University, where they were informed that the persistent "noise" was probably the most important radio signal ever to be received from outer space. This 2.7 K radiation, identified as the afterglow of the Big Bang, suffuses the sky in all directions at microwave frequencies.

The cosmic microwave background (CMB) of 2.7 K radiation was interpreted as providing direct evidence of how radiation was distributed throughout the Universe when it was less than one million years old. In 1989 the first Cosmic Background Explorer (COBE) brought back recordings of a uniformity not varying more than 1 part in 100,000 in all directions. Advocates of the BB theory immediately interpreted these results, along with the thermal nature of the cosmic microwave background as evidence of its primodial origin.

In the same 1992 article, Burbidge wrote, "The pervasive cosmic microwave background was predicted by the big bang theory and is still considered to be one of its strongest pieces of supportive evidence. Measurements now, however, show that the background radiation is extremely smoothly distributed. Maps of galaxies, on the other hand, show structure on all scales."

"According to the standard version of the big bang theory, matter and radiation were strongly coupled together in the early universe, and only later did the two go their separate ways. If this were so, the cosmic microwave background would show some imprint from the lumpy matter distribution that led to the formation of galaxies. In actuality, however, the cosmic microwave background appears smooth to at least one part in 100,000, so close to the level at which the big bang must be abandoned or significantly modified."

But not to worry: The COBE Satellite had been sent aloft again, and recorded another cosmic microwave background of 2.7 K radiation. Just two months after the Burbidge publication, the interpretation of this COBE probe became public. Incredulously, the presence of minuscule ripples -- small temperature fluctuations of slightly more than 1 part in 1 million -- was announced, and advocates of the BB breathed a huge sigh of relief. The tiny ripples were interpreted as fluctuations in the density of matter and energy in a very early phase of cosmic history: the clumping of matter into larger structures such as galaxies.

In the same article, Burbidge had written, "Within the framework of the hot big bang, there is no satisfactory theory of how galaxies and larger structures formed. Galaxies cannot form by gravitational collapse in an expanding universe unless one assumes without explanation that large density fluctuations were present in the early universe. Under the influence of particle physicists, cosmologists are now proposing that these fluctuations occurred at an early stage of the big bang or else were caused by exotic entities such as cosmic strings. None of these ideas can be directly tested."

"The inflationary model, a pet idea of the past decade, holds that a period of extremely rapid expansion in the early universe accounts for both the smoothness of the cosmic microwave background and for the amount of matter present in the universe. But again, inflation is an untestable addition to the lore of the big bang."

Further, if "a period of extremely rapid expansion in the early universe" did happen, wouldn't this disrupt calculations of extrapolations backward to the time of the BB? To quote Burbidge again, "This form of inflation is arbitrary, and our successors will wonder when it goes out of favor, as the history of science suggests it will, why is was so popular?"

Whether or not the new finding of minuscule fluctuations in the background radiation eliminated the necessity of the inflationary model is not clear in many minds. But when we take a closer look at the CMB radiation, the question becomes moot.

A CLOSER LOOK AT THE BACKGROUND RADIATION

While the cosmic microwave background is identified as a relic of the big event, the true source of the 2.7 K radiation remains debatable. The major problem with this pervasive radiation resides in its interpretation. In reality, it is blackbody radiation, and such radiation is characteristic of all Solar System objects. (Ref: *Encyclopedia of Physics*).

Consider the radiation in the space enclosed within a hollow steel ball which is maintained at a constant white heat. The radiation within such a sphere is called "blackbody" radiation because the radiation coming from each unit area of the walls of the enclosure is the same as that which comes from unit area of a perfectly blackbody maintained at the same temperature. The radiation from each unit area of the walls of the enclosure is determined only by the temperature of the walls and not by the substance of which the walls are composed. In such an enclosure the intensity of the radiation falling on a unit area of the walls equals the radiation coming from the unit area.

In principle, the same thing applies to a planetary sphere in which the white heat equates to its nuclear energy core and the mantle/crust equates to the walls. *The radiated energy originates from the internal energy associated with atomic and molecular motion and the accompanying accelerations of electrical charges within the object (sphere).*

Scientists realized that the results of the discovery of Penzias and Wilson were limited to only a few wavelengths clustered at one end of the Planck curve. Other explanations of the background radiation, such as a combination of radio sources, could explain those data points. It was not until the mid-1970s that enough measurements at different frequencies had been made to prove that the background radiation actually follows Planck's law.

Radiation is both absorbed and emitted by the walls of the enclosure, and by any gas which might be in the enclosure. Planck suggested the idea that when radiation is emitted or absorbed, it is emitted or absorbed in multiples of a definite amount, identified as a quantum. The energy of a quantum of radiation of frequency v is proportional to v and so is equal to hv, where h is a proportionality constant known as Planck's constant.

The full curve of Planck's law is the spectrum of the continuous radiation given out by a hot blackbody. The plotted curve shows the distribution function with respect to wave-length and/or frequency. It applies to any planetary mass containing a hot nuclear core.

No problem would exist here if scientists could accept the fact that the spheres of the SS are evolving from nuclear masses, and the CBR is in reality the radiation emanating from these internal nuclear reactions. Simply put, the relationship among the three characteristics of these internal nuclear reactions is best shown by the full curve of Planck's law: the spectrum of the continuous radiation given out by these hot black bodies.

Proponents of the BB theory claim evidence that the CBR originated at extremely high redshift. The evidence comes from the production of light elements. Knowing the present CBR temperature, and assuming entropy was nearly conserved back to high redshifts, they trace the thermal history of the Universe back in time to temperatures high enough to have driven thermonuclear reactions. Computations of the nuclear reactions, under three other (conditional) assumptions, predict that most matter comes out of the "hot BB" as hydrogen, with about 20% by mass helium, and significant amounts of deuterium, ^{3}He and lithium. The computation, when using a mean baryon number consistent with what is observed [supposedly] to be present in and around galaxies, yields computed abundances *concordant with what is seen in old stars.*

Utilizing the same information except for the assumptions necessary in the BB theory, similar reasoning in the perspective of the FLPM/IN concept allows one to reach another confirmatory conclusion. Rather than looking to the BB for answers to the origin of light elements, the computation of the nuclear reactions can be applied most assuredly to the nucleosynthesis common to all planets and stars. No complicated assumptions are necessary. As shown in previous chapters, the production of all elements therein will be in the same proportions predicted in the calculations used in the BB perspective.

Planck's law applies to all thermonuclear reactions, no matter in which theory it is utilized. The above information ranks high among the many crucial clues that reveal the true nature of planetary origins. *While negating any claim that the CBR exclusively supports the BB theory, the facts stand firmly as powerful confirmatory evidence of the blackbody nature of the spheres of our SS -- the true source of the cosmic background radiation.*

Consequently, any COBE search within the realm of our SS cannot escape the effects of the blackbody radiation emanating from objects comprising the system. Isn't it reasonable then to conclude that no matter where Penzias and Wilson pointed their microwave horn antenna, they inevitably would record the same noisy emissions in every direction?

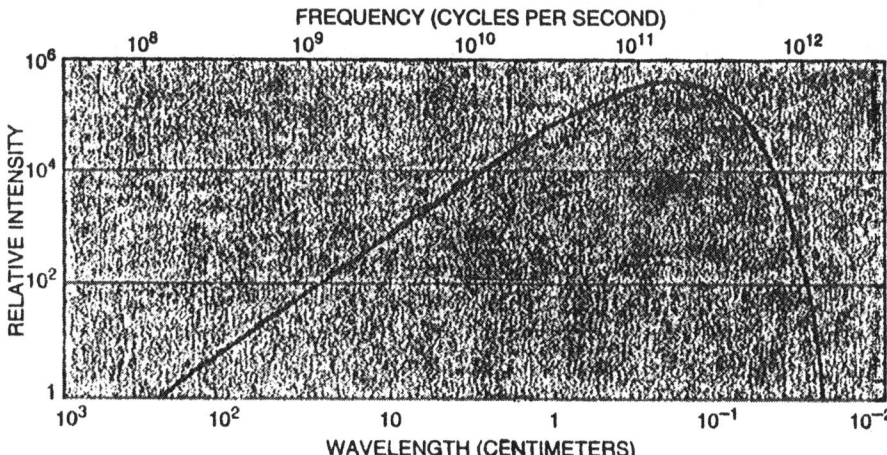

Planck's curve of blackbody spectrum applies to all thermonuclear reactions. It is powerful confirmatory evidence of the blackbody radiation emanating from the spheres of our Solar System -- the true source of the cosmic background radiation recorded in every direction.

Even a probe escaping the effects of our SS would never be able to escape the "noise" effects of other suns (stars) undergoing transformation in the normal course of evolution. Starting within our own planet, it can be heard as the "noise" of nucleosynthesis -- the hum of creation -- that does indeed pervade throughout the Universe. As long as there is an active, expanding Universe, the hum will not cease.

Certainly, the eternal *Hum of Creation* is a suitable name for this 2.7 K blackbody radiation. Rather than a remnant of the BB, it is the humming of ongoing creation by which our planets evolve through the five stages of evolution and our Sun slowly sacrifices its mass to make the system possible.

But keep in mind that this persistent hum of the spheres bears little relevancy to Kepler's long-sought, elusive music of the spheres. That tune dwells within the beautiful Phi geometry of the SS.

Misinterpretation of the significance of the 2.7 K radiation is a prime example of many discoveries that have been force-fitted into the BB theory in efforts to advance the cause of a concept already fatally flawed. An excerpt from a letter published in *Science News* (27 Jul 1991) reads: "When the Big Bang proponents make assertions such as 'a whole bunch of observations that hang together', they overlook [misinterpret contradictory] observational facts that have been piling up for 25 [now 30] years and that now have become overwhelming."

Certainly, when this new interpretation of the persistent noise of creation is accepted in the scientific community, it will be a powerful thrust into the heart of the BB. In the following section, other misinterpretations of observed phenomenon will be examined.

THE RELATIVE ABUNDANCE OF ELEMENTS VIA THE BB

The seed for Big Bang cosmology was planted in the 1930s, after Edwin P. Hubble, an American astronomer, discovered that galaxies recede from one another and that the most distant ones recede at the greatest rates. Hubble's discovery of this ongoing expansion was interpreted to imply that the cosmos had once been concentrated as a very small unit that exploded into our universal system.

To make the system work, scientists had to start with a tiny unit of mass at a temperature measured in trillions of trillions degrees. The tiny unit exploded instantaneously, spreading in all directions as a hot cloud of energy. Nuclear physics provided the tools for modeling the synthesis of the elements from fundamental particles within this hot cauldron.

The initial quantitative theory was developed by George Gamow, a Russian-born physicist, who had built a reputation by explaining radioactive decay.

In the early 1930s, scientists realized that most stars consist of hydrogen and helium. Since the hydrogen nucleus contains only one proton, it seemed reasonable to assume that it was the first element to form, or precipitate, from the hot expansion as it began cooling down. Helium, consisting of a nucleus containing two protons and two neutrons, was next in the line of elemental formation, created by the fusion of hydrogen.

But the fusion of protons requires powerful forces to overcome the immense electrostatic repulsion between them. Scientists believed that the tremendous heat and pressure required in this process could be furnished only by a primodial event or the interior of a star. The prevailing theory of the nuclear physics of stars had been established by Hans Bethe in his explanation of how the Sun shines: nuclear fusion in stellar interiors converts mass into energy. This led to Hoyle's solution in which carbon is produced from three helium nuclei under the severe conditions in star cores.

By 1957 a scheme explaining how stars might have synthesized most of the elements from hydrogen and helium had been worked out by Fowler, Hoyle, Margaret Burbidge and Geoffrey Burbidge. The work was done independently by A.G.W. Cameron in Canada.

But the cosmic abundance of helium remained a mystery, which was eventually solved by Gamow. He suggested that the elements might have formed in an extremely hot, dense gas of neutrons, even before the stars came into existence. Some of the neutrons decayed into protons and electrons -- the building blocks of hydrogen. According to this theory, larger nuclei formed in the primeval inferno when smaller ones, beginning with hydrogen, grew through the successive capture of neutrons. This process continued until the supply of neutrons was exhausted, the temperature dropped and the particles dispersed. Hoyle has the distinction of calling this the big bang theory, a concept of creation of the elements that prevails today.

Initially, the BB theory failed to explain the formation of the elements beyond helium, which has a mass number of four. Because there are no stable isotopes having mass numbers of five and eight, elements heavier than helium cannot be made by adding neutrons to it one at a time. This problem was solved only by invoking the stellar nucleosynthesis of Hoyle, Fowler and their associates. Therefore, the heavier elements had to have been made in stars and supernovae.

Thus, the origin and the abundance of the elements found throughout the

Universe are supposedly explained by the BB theory.

In a strange twist of fate, the BB theory has proven to be an important step in confirming the evolutionary process in which elements are also made in planets via nucleosynthesis. While evidence mounts against the BB theory, it grows more overwhelmingly in favor of the Little Bangs theory of the late 1970s. While it is true that elements are made via nucleosynthesis in stars, our Sun and in supernovae explosions, scientists have somehow managed to over-look the most obvious source of creation of the elements: the nucleosynthesis in SS spheres.

Here the BB can take much of the blame, which, in turn, can be passed back to the Kant-LaPlace era in which the hypothesis of the formation of our Solar System from a gaseous dust-cloud originated. In this perspective, scientists found it necessary to have a source of this material. Here the BB theory in which clouds were dominant seemed to fit right in. This, in turn, has led to erroneous conclusions that all universal spheres formed via condensation processes. Evidence shows that nothing could be farther from the truth.

Quoting Burbidge again: "Big bang cosmology is probably as widely believed as has been any theory of the universe in the history of Western civilization. It rests, however, on many untested, and in some cases untestable, assumptions. Indeed, big bang cosmology has become a bandwagon of thought that reflects faith as much as objective truth."

He closes with this mutual thought: "Why then has the big bang become so deeply entrenched in modern thought? Everything evolves as a function of time except for the laws of physics. Hence, there are two immutables: the act of creation and the laws of physics, which spring forth fully fashioned from that act. The big bang ultimately reflects some cosmologists' search for creation and for a beginning. That search properly lies in the realm of metaphysics, not science."

THE FIERY LITTLE BANGS THEORY: A Plausible Alternative

In 1993, the author found a copy of Einstein's Universe (1979) in which Nigel Calder wrote an opinion on the BB theory: "It may be that no one has yet thought of a better and truer scheme for the Universe." At that time, Calder was among the vast majority who were unaware that a more logical scheme had begun to take shape in the author's mind in 1973 and was being stubbornly pursued, persistently developed, and to a very limited degree, recorded in the scientific literature. The Little Bangs theory of 1979 became a vital link in the FLPM/IN concept that finally came to full fruition in 1995 when the last Diagrams of the SS evolved into the Fourth Law of Planetary Motion as the final piece of the giant puzzle of our origins.

Even at this writing, new discoveries continue to add more and more exciting evidence, each bit driving another nail into the coffin of the BB theory while adding substantial support to a Little Bangs concept that appeared as early as 1979 to tie everything together. Calder's book seemed to vindicate my findings of the previous 20 years, while giving incentive to push onward.

The initial clue to the Little Bangs theory (LBT) was rooted in Hubble's discovery that galaxies are moving rapidly away from us and from each other, and the farther away they are, the faster they are moving. The prevailing BB interpretation that an explosive force imparted such vast momentum to the matter now comprising the distant galaxies did not make sense, especially since they were and still are moving aggressively, after billions of years, against strong forces of a centralized gravity.

Further, astronomers now know that galaxies at great distances are significantly younger than nearby galaxies. And on balance, present evidence favors unending expansion (not possible in the BB version) and the dire necessity of a continuity in nature.

In my mind, the utilization of powerful forces at the spherical perimeter of the expanding Universe was essential in keeping the system functioning in a perpetual motion manner. Black holes and quasars could be the sources of the powerful energy so essential to the system. The concept fell together in 1979.

As the white energy of light reaches the spherical Universe at its perimeter, it is slowed somewhat, and collapses into the form of tiny black holes, which continue outward at nearly the speed of light while continuing to gather energy also from the darkness of the space it enters. Almost immediately, a black hole becomes a quasar, shining brilliantly because of the subsequent in-falling light crashing violently and noisily onto its perimeter.

It seemed reasonable to believe that the speed of the in-falling light increases dramatically as it is pulled into the new energy mass -- perhaps attaining a velocity as high as the c^2 speed figure in Einstein's famous formula, $E = mc^2$.

Perhaps this formula is telling us that energy is created by the in-falling mass traveling at the speed of light squared. If so, the relevant mass can be determined by dividing E by c^2.

In the same book, Calder wrote: "Physicists working with atomic particles found that light of sufficient energy -- very energetic 'gamma-rays' to be precise -- could make fresh atomic particles. The energy of light was transformed into matter." At the time, I was unaware of both Calder's book and the matter-antimatter reaction. By sheer coincidence, we had reached essentially the same conclusion through similar processes of thought during the same year. The LBT formulated at that time predicted that light energy exiting at the perimeter of the spherical Universe coalesces into 'black hole' energy that later forms into bright, noisy quasars that eventually erupt into a series of galaxies, thereby explaining the speed-of-light expansion of the Universe (galaxies and quasars moving at that speed at the perimeter).

Calder continued the coincidence: "When a black hole is continuously fed with matter, its surroundings can glow extremely brightly. The falling matter exudes the energy as if in a dying shriek, before it disappears forever. ...quasars are the small cores of violently exploding galaxies, far outshining the ordinary stars."

This description was very similar to my description of a quasar in describing its formation from a black hole, and later exploding into a series of galaxies -- written the same year, but only discovered 14 years later. It served as a vital part of the Little Bangs theory of 1979. The manuscript was copyrighted, but unfortunately, no editor accepted it for publication. It remains a good dust collector.

Quoting Calder again "...when a particle meets its anti-particle, they annihilate one another and disappear from the Universe. All that remains is a 'puff' of gamma-rays -- a reversal of the process of creation."

Further, "All that is required to create a black hole ...is that light should feel the effects of gravity. ...space cannot exist separately from 'what fills space', and the geometry of space is determined by the matter it contains."

At this time, it would be worthwhile to consider another, but similar opinion. Edward Tryon, City University of New York, has boldly stated that the Universe could have been created out of absolutely nothing without violating any of our physical laws, if it has certain properties. He wrote, "Every phenomenon that could happen in principle actually does happen occasionally in practice, on a statistically random basis. For example, quantum electrodynamics reveals that an electron, positron and photon occasionally emerge spontaneously from a perfect vacuum. When this happens, the three particles exist for a brief time, and then annihilate each other, leaving no trace behind. ...[This] is called a vacuum fluctuation, and is utterly commonplace in quantum field theory."

Continuing with the LBT: after ages of time, the brilliant quasar explodes, sending brilliant fireballs of energy embedded in clouds of energy-matter that spread in whirling finger-like formations. Out of this fiery soup a group of galaxies are created in a spherical bubble-shaped formation around the central point of the explosion. Such bubble-shaped formations are observable in the distant sky.

Closer examination shows a clear hierarchy of galaxies grouped into clusters, and clusters into superclusters. Such formations do not seem possible in the time frame and perspective of the BB theory -- in spite of the tiny fluctuations in the COBE recording.

Within each whirling galaxy, the fireballs react with the cooler enveloping clouds. Fiery masses near enough to each other interact to form binary star systems, a common sight to astronomers. When some of these systems meet the right conditions of relative sizes, velocities and distance apart, the result can be a solar system like the three recently discovered: 51 Pegasi, 70 Virginis and 47 Ursae Majoris. In very rare cases, they go a step further and meet the special prerequisites that cause the smaller, faster mass to break up at the points dictated by Nature's Phi geometry (Chapter I). The result is a SS like ours.

UNIVERSAL ENERGY: A BIBLICAL CONNECTION?
(*From Void to Energy to Universe*, 1980, excerpt)

While the Universe appears to be creating energy, according to the Expansion-by-Quasar Theory, it is possible that mankind will be able to create energy some day in similar manner. Much has been accomplished already in the field of near-absolute temperatures, and man has discovered things believed to be impossible. The possibility that we may some day learn to create energy by duplicating Nature's, or God's, process will always haunt man until he does accomplish it.

At this point, let's take another look at the Biblical Book of Genesis, according to the Authorized King James Version. It states, "In the beginning God created the heaven and the earth. And the earth was without form, and void; and darkness was upon the face of the deep. And the Spirit of God moved upon the face of the *waters (void?). And God said, Let there be light; and there was light. And God saw the light, that it was good: and God divided the light from the darkness."

The important key that relates Creation according to the Biblical version, to Creation according to the Expansion-by-Quasar Theory, is the emphasis placed upon Light. The Bible does appear to be telling us that Light was the key factor in Creation. It was this Light that started the Universe on its course. There are no scientific explanations, as far as I have been able to determine, for the origin of such Light and its subsequent Energy Creation. The only remaining question of pure logic is: Did God make Himself into pure Light and Energy, knowing its potential as a Universe, and foreseeing, planning such a complicated, yet comprehensible system? Was Man created to eventually solve all mysteries of the Universe — even the question of God's real purpose for Man in the Universe? My own answers, in all cases, always lie on the affirmative side of the issues.

The pure physicist will, for some time to come, remain adamant in his adherence to the law of conservation of energy: Mass-energy cannot be created or destroyed, but each may be converted into the other. And understandably so, for this has been a sacred law of physical chemistry since Lavoisier wrote the first modern textbook of chemistry in 1789.

Certainly, the law of conservation of energy holds true within the limits of knowledge and capabilities of mankind. But what happens out beyond our limitations? It seems an almost irrefutable fact that mass-energy could not exist if it had not been created, period. Mankind's destiny, as throughout history, always lies beyond his limitations, but only temporarily, in each specific situation. That man will learn how to create energy as Nature does is perhaps, inevitable.

MORE ON GALAXIES

Among the recent findings of the Space Telescope was a distant galaxy billions of light years away that looked too modern at a time when the Universe was supposedly a tenth of its present age. Hubble images show that the spiral shape of galaxies was far more common among these youthful galaxies than in ones closer to us in space and time. This was surprising, because many astronomers had postulated that ellipticals would form more slowly than spirals. And they had formed so much earlier than astronomers expected galaxies of any kind to form.

Yet, going back 12 billion years ago, nine-tenths of the way back to the BB, Hubble had photographed a fully formed galaxy. When Duccio Macchetto of the Space Telescope Science Institute in Chile aimed the Space Telescope at this mass in the sky, he saw the unmistakable image of an elliptical object.

Proponents of the BB do not believe that the presence of the full-fledged elliptical so soon after the Universe is supposed to have formed casts doubt on the reality of the BB, but it does raise questions about the true nature of the beginning. One favored variant of the BB theory assumes the spawning of a Universe containing a density of matter high enough to eventually halt its expansion. But that model also predicts that primodial matter was distributed so evenly that galaxy formation was delayed. Consequently, Macchetto's discovery may force cosmologists into a version of a less dense, ever-expanding, open universe, in which galaxies could take shape earlier. If so, it will be another giant step in the direction of the Little Bangs concept.

In *Asimov on Astronomy* (1974), Asimov made the following prediction: "In fact, even in a finite Universe, with a radius of 12 billion light years, there might still be an infinite number of galaxies; almost all of them (paper-thin) existing in the outermost few miles of the Universe-sphere."

This prediction might not be too far off. The photograph made by the HST in December 1995 is the deepest archaeological dig in the history of galaxies. The great abundance of faint galaxies some five-sixths of the way to the edge of the visible Universe presents a problem. Much interest will focus on the 1500 to 2000 bluish dots, which many astronomers believe are young galaxies in the distant Universe. The implications extend well beyond one tiny patch of sky. Based on it, a survey of the entire sky to the same depth would reveal a total of 50 billion faint objects. To make such a map would require a million years.

At these great distances, most astronomers expected galaxy numbers to decrease. The fact that they don't decrease expands the population of known galaxies and leaves an uncomfortably short time for them to form after the BB. Further, radio galaxies and quasars are more densely packed at distant phases of the Universe than near at hand. Here again, the evidence mounts against the BB theory, while adding powerful support to the LB theory of expansion and growth of the Universe at its spherical perimeter. Not surprisingly, all things in

Nature grow in similar manner; e.g., a tree grows upward and outward by adding new matter to the ends of all the branches.

The immensity of the known Universe continues to grow by leaps and bounds. These vast quantities begin to cast more doubt that they all could have been stored inside a pinhead. Simultaneously, they add significant credibility to the Little Bangs theory in which they are created continuously.

An interesting comment on the BB theory was made recently in the weekly *Ask Marilyn* column. Responding to an inquiry, she wrote: "I think that if it had been a religion that first maintained the notion that all matter in the entire universe had once been contained in an area smaller than the point of a pen, scientists probably would have laughed at the idea."

In expressing his cynical view of science, Planck also gave hope for the future. Writing in his *Scientific Autobiography and Other Papers* (1949), the great physicist argued, "A new scientific truth does not triumph by convincing its opponents and making them see the light, but rather because its opponents eventually die, and a new generation grows up that is familiar with it."

HOW STARS FORM

On April 1, 1995, the Hubble Space Telescope snapped a photograph showing about 50 stars inside monstrous columns of dust and especially dense molecular hydrogen gas. The gaseous towers, six trillion miles long, resemble stalagmites rising from a cavern floor. At their edges can be seen finger-like protrusions, each with tips larger than our SS, in which the stars are embedded.

The information released to the public six months later interpreted the photograph in the perspective of the BB theory: the stars supposedly are created when "the gas collapses under its own gravity." As they grow, the massive stars produce huge amounts of intense ultraviolet radiation that hollows out a cavity around them by heating their surroundings, thereby making them visible. As the cloud gets boiled away, it uncovers stars buried there.

Interpreted in the Little Bangs perspective, the monstrous columns of gas and fiery stars are products of a tremendous explosion (or a series of explosions) that imparts the momentum essential for interactions that can result in active binary systems while scattering them eventually over the six-trillion-mile distance. The stars were created as fireballs in the explosion(s), rather than

124

from the clouds in which many still remain hidden from view. However, the two viewpoints do agree on the cavity formation aspect of the theory.

SOLUTIONS TO ANOMALIES: A SUMMARY LIST

The ramifications of the geometric solution to the origin of the SS are immense. As the final phase of the FLPM/IN concept, it completes the continuity of evidence that weaves together all the anomalies of our SS. Most of the anomalies listed below have been discussed in this book or in previous writings. All are either solved or solvable in this new perspective.

1. Why planets obey Kepler's First Three Laws of Planetary Motion.
2. The complete geometry of the SS.
3. The original and current spacing of planets.
4. The valid explanation of the enigmatic Bode's Law of spacing of planets.
5. Why Neptune and Pluto do not obey Bode's Law.
6. Why planetary orbits are inclined to the ecliptic.
7. Why the displacement (AU) of each planet from its original orbit.
8. The highly elliptical and large inclination of the orbit of Pluto.
9. The speeds and directions of rotation and revolution of planets.
10. The slow reverse spin of Venus.
11. The huge size of Jupiter (the Geometric Mean of the SS).
12. The tapered sizes of the planets.
13. How large moons formed contemporaneously with planets.
14. Why the Sun's equatorial mass rotates faster than the remaining mass.
15. Why the Sun's equatorial mass has a 7° inclination to the ecliptic.
16. The huge discrepancy in angular momentum of the Sun and the planets.
17. The nature of Earth's core: nuclear energy (ten clues).
18. The abiogenic origin and evolution of hydrocarbon fuels (1973).
19. The origin of surface features: water, land, salt mines, etc.
20. The expanding Earth: increases in sea level and land.
21. How the Asteroids came into being: Olbers was right.
22. The craters on moons, asteroids, comets and planets: Mills was right.
23. How planets evolve through five common stages in accordance with size..
24. The physical and chemical differences among planets and moons.

25. Why species come and go: the extinction of the dinosaurs

26. Why Earth and other planets are self-sustaining entities.

27. Plate tectonics: from highly mobile to barely movable.

28. Why the electromagnetic strength of each planet is different.

29. The unexpectedly powerful explosions of Comet SL9 on Jupiter.

30. The origin and evolution of comets.

31. The origin and fate of planetary rings.

32. The cometary moons of Mars: Phobos and Deimos.

33. How the discoveries of the space probes fit into the FLPM/IN concept.

34. The 2.7 K radiation throughout the Universe: not from the Big Bang.

35. The erroneous notions of the Big Bang.

36. The too-high Hubble constant revealed by the Hubble Space Telescope.

37. The new conflict between the oldest stars and a younger Universe.

38. How and why the Universe expands at its spherical perimeter -- forever.

39. The 51 Pegasi system discovered by the HST.

40. The large extrasolar planets: 70 Virginis and 47 Ursae Majoris.

The full significance of the Fourth Law of Planetary Motion yet lies camouflaged within the mysteries of the SS and among the enigmas of the Universe.

Einstein was dogmatic about continuity in Nature, and rightly so. Nigel Calder wrote, "The uncomprehending antagonism evoked by Einstein's ideas is a sign that the old conflict between scientific inquiry and dogma is far from dead." Perhaps it always will be so.

Quoting Calder again: "...great syntheses are made in individual minds." [Not by committees, as has been shown many times in the past]. "...all that is lacking is new comprehensive insight... That must come soon."

To the credit of science, the majority of scientists do not believe the Big Bang theory. But as long as the concept prevails in the news media, any new insight will have a tough uphill battle just to establish a foothold with that majority. At this writing, the new FLPM/IN concept has not proven to be an exception to the rule.

One day in September, 1992, in frustration of the situation, I wrote: "Even as the 21st century approaches, the science of planetary origins and evolution remains bogged in a quagmire of egoistic bias, snugly entrenched in an ivory tower built of cards on a foundation of sand, with shades drawn against the light of new ideas and factual knowledge, reveling in the fantasies of hypotheses that would burst like a bubble with the singular probing of the finger of question."

While it did relieve the frustration somewhat, the situation has shown no perceptible sign of change. Perhaps some day the system will improve, but right now there is no light at the end of the tunnel.

Perhaps this forced self-publication will add enough substance to the already overwhelming evidence against current dogma to force a change in the direction of scientific thought on the origins of our SS and the evolution of its planetary systems. In view of the continuity of evidence herein, *certainly the Little Bangs theory is a plausible alternative to the Big Bang theory. The great beauty of it is that no changes in established evidence should be necessary; however, it is essential, even crucial, to the advancement of the planetary sciences and astronomy that the perspective in which this evidence is interpreted be examined closely.*

In a battle for truth, one cannot afford to tread too lightly -- or too heavily.

Great Comet of 1861
Note the 8-tail ejecta from 8 crater ports.

Morehouse's Comet
Note the multi-port ejecta pattern.

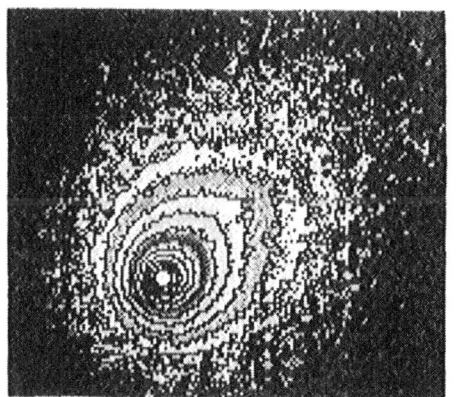

A typical false-color image that is always identical for comets and background stars.

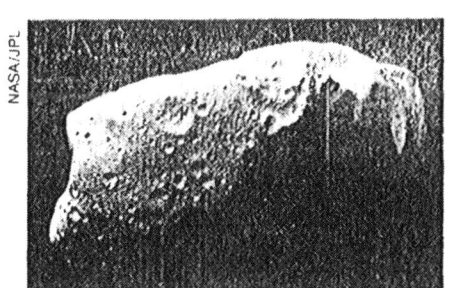

Galileo's view of the cratered asteroid
243 Ida.

www.ingramcontent.com/pod-product-compliance
Lightning Source LLC
Chambersburg PA
CBHW081152180526
45170CB00006B/2034